**NONLINEAR SCIENCE** THEORY AND APPLICATIONS

Numerical experiments over the last thirty years have revealed that simple nonlinear systems can have surprising and complicated behaviours. Nonlinear phenomena include waves that behave as particles, deterministic equations having irregular, unpredictable solutions, and the formation of spatial structures from an isotropic medium.

The applied mathematics of nonlinear phenomena has provided metaphors and models for a variety of physical processes: solitons have been described in biological macromolecules as well as in hydrodynamic systems; irregular activity that has been identified with chaos has been observed in continuously stirred chemical flow reactors as well as in convecting fluids; nonlinear reaction diffusion systems have been used to account for the formation of spatial patterns in homogeneous chemical systems as well as biological morphogenesis; and discrete-time and discrete-space nonlinear systems (cellular automata) provide metaphors for processes ranging from the microworld of particle physics to patterned activity in computing neural and self-replicating genetic systems.

*Nonlinear Science: Theory and Applications* will deal with all areas of nonlinear science – its mathematics, methods and applications in the biological, chemical, engineering and physical sciences.

**Nonlinear science: theory and applications**

*Series editor:* Arun V. Holden, *Reader in General Physiology, Centre for Nonlinear Studies, The University, Leeds LS2 9NQ, UK*

*Editors:* S. I. Amari (Tokyo), P. L. Christiansen (Lyngby), D. G. Crighton (Cambridge), R. H. G. Helleman (Twente), D. Rand (Warwick), J. C. Roux (Bordeaux)

**Chaos**

*Editor:* A. V. Holden

Other volumes are in preparation

# Control and optimization

## The linear treatment of nonlinear problems

J. E. Rubio

*School of Mathematics*
*University of Leeds*

Manchester University Press

Published by
Manchester University Press
Oxford Road, Manchester M13 9PL, UK
*and*
51 Washington Street, Dover,
New Hampshire 03820, USA

*British Library cataloguing in publication data*
Rubio, J. E.
Control and optimization: the linear treatment of nonlinear problems. – (Nonlinear science, Theory and applications)
1. Control theory 2. Mathematical optimization
I. Title II. Series
629.8′312 QA402.3

*Library of Congress cataloging in publication data*
Rubio, J. E.
Control and optimization.
Bibliography: p. 118
1. Control theory. 2. Mathematical optimization.
I. Title.
QA402.3.R83 1985 629.8′312 85-15256

ISBN 07190 1841 2 *cased*

Typeset in Hong Kong
by Graphicraft Typesetters Ltd
Printed and bound in Great Britain
by Biddles Ltd, Guildford and King's Lynn

# Contents

# Preface

This book presents the results of some of my researches on optimal control theory, carried out over the last decade. Motivated by a book on the calculus of variations by L. C. Young, I saw some time in the early seventies a way of treating optimal control problems which could have some advantages over the usual, classical approach. This formulation is based on the injection of the classical admissible trajectory – control pairs into a space of measures; the classical problem is replaced by a measure-theoretical one, in which one seeks to minimize a linear form over a subset of this space determined by the constraints of the problem. In fact, this subset – of positive measures – is described by linear equalities; it is possible then to use the whole paraphernalia of linear analysis to treat seemingly nonlinear problems.

The first four chapters present an exhaustive study of the problem mentioned above. After giving an introduction to the present approach in Chapter 1, together with an overview of the main results presented in the rest of the book, existence theory is studied in Chapter 2; several theorems are proved, under different hypotheses. The role of linearity is emphasized in Chapter 3, in which the measure-theoretical optimization problem is treated as an infinite-dimensional linear programming problem. Of course, it is necessary to approximate the action of an optimal measure by that of trajectory – control pairs; a sequence of such pairs is constructed in Chapter 4, on approximations. In Chapter 7, the theory is extended to the case in which the state space of the controlled system is a Hilbert space.

Chapters 5 and 6 are concerned with applications of the theory developed in the rest of the book. A computational method is developed in Chapter 5, which appears to be adequate for a variety of optimal control problems; an interesting by-product of this is a method for the global optimization of a real function of several real variables. Global nonlinear controllability is treated in Chapter 6; two different approaches to this subject, based of course on the measure-theoretical formulation, are explored.

Chapter 8 deals with several aspects of the control of the diffusion equation, from a measure-theoretical standpoint related to the one used in previous chapters in connection with finite-dimensional systems. When studying the controllability of the diffusion equation, however, it was found necessary to go beyond measures – indeed, beyond distributions! The research work which gave rise to this chapter was a joint effort with Dr D. A. Wilson, of the Electrical and Electronic Engineering Department here at Leeds.

When planning this book, I had to decide whether to include some mathematical background. In the end, I decided to include an appendix which gave an overview of the connections between classical measures and linear functionals, and a detailed proof of Riesz's theorem, which is seldom found in easily accessible textbooks. Also, I have included a thorough treatment of the structural theory of the extremal points of a set of positive measures defined by linear equalities. Needless to say, this appendix is not a substitute for the several books which a beginner should master; I would recommend that beginner to read and learn well most of the book by TRÈVES [1], most of the first volume of the treatise by CHOQUET [1], some relevant chapters in BERGE [1], and, if a good introduction to somewhat more elementary analysis and measure theory is needed, most of the material in FRIEDMAN [1].

# Acknowledgements

Many people helped me in the writing of this book, in many different ways. Professor H. L. Price, of the Department of Applied Mathematical Studies here at Leeds, made this book possible by actively encouraging the development of control-theoretical studies in the Department. I would like to thank Professor F. A. Goldsworthy, Head of the Department, for rearranging my teaching schedule so that I could spend a whole term concentrated mostly in writing the book. My colleagues, Mr David Knapp and Mr Ed Redfern have helped, especially when guiding me through the intricacies of the operating system of our computer. We have had many interesting discussions with Dr Sam Falle in matters of computation, mainly in connection with the optimization method presented in Chapter 5.

Special acknowledgements and thanks should be given to three control scientists:

Dr David Wilson, from the Department of Electrical and Electronic Engineering here at Leeds, made many contributions to the research leading to the results reported in Chapter 8, and indeed the material in the first half of that chapter was developed almost exclusively by him.

Dr Richard Vinter, from the Department of Electrical Engineering of Imperial College, London University, after being introduced to this theory in a lecture I gave at Imperial College, has himself made many distinguished contributions in this field, in a somewhat modified framework, especially with the development of necessary and sufficient conditions for optimality; he has given me counsel and advice over the years.

Dr S. P. Banks, of the Department of Control Engineering of Sheffield University, read the manuscript in detail.

Dr Arun Holden, Editor of this Series, made a welcome invitation to submit the manuscript for possible publication in it, and has given me constant support; both he and Mr Alec McAulay, of the Manchester University Press, have been very helpful solving the problems which arose in the production of the book.

This book is dedicated to the memory of the late RICHARD BELLMAN, whose idea it was; it is doubtful whether it would have been written without his advice and encouragement over the years.

# 1 Optimal control problems, old and new

## 1 Introduction

Those were heady days, then in the sixties, for control theorists. The application of mathematical methods, sometimes borrowed from that old discipline, the calculus of variations, had paid off handsomely, and everybody was talking about Pontryagin's principle, dynamic programming, controllability (and, sometimes, about *complete* controllability!), and the new computational methods.

So time went by. New results have indeed been forthcoming. The theory of necessary conditions has been developed successfully; existence theory is comprehensive and sophisticated; the control of systems described by partial differential equations has been exhaustively studied, and there have been many advances in optimization theory.

Why is it then that there is, at least for this writer, a sense of disappointment, a feeling of the morning after. No doubt that the pace of innovation has slowed down; we are in a situation not uncommon in scientific or mathematical fields, a period of consolidation following an era of change and progress. It is unfortunate that so many problems are left to be solved; it is not really an appropriate time for consolidation! Those sophisticated necessary conditions rarely give an insight into the structure of the optimal controls. No really comprehensive method for the numerical estimation of optimal controls seems to have emerged after all these years. One would like to understand nonlinear controllability.

One unfortunate consequence of consolidation has been the growth of an orthodoxy, the choice of models and methods being dictated sometimes – often – by familiarity, tradition, deference. Some years ago we wrote with one of our students a paper, an application of control theory to mathematical economics; we used an integral performance criterion in which the integrand was the absolute value of some variable. The consequences of this choice were rather surprising and certainly unorthodox – that the best way to manage an economy was to go from stop to go, from bust to boom, in a manner which can loosely be described as infinitely often. (This seemed a fairly good description of British economic policy at the time; it appeared that successive governments had stumbled onto our prescription.) The reviewer was indignant. *Performance criteria should be strictly convex.* (He meant, we should have used the square of the variable as an integrand.) What were we up to, using a merely convex criterion? We answered back, why a strictly convex criterion? What is wrong with mere convexity? The answer was no more than, well, everybody uses squares, one can understand the results, the great Bloggs uses them, who are you to do otherwise.

Of course there are many works that point in other directions (see, for instance, WARGA [1] and IOFFE and TIHOMIROV [1]). We would be satisfied if this book were to make a contribution to the movement away from orthodoxy initiated by these authors; it is very necessary to suggest new vistas and possibilities. After all, not every plant is linear, not every performance criterion quadratic. We are not providing in this book an instant, magic replacement for the old workhorses; we are planting a seed which may give rise to some useful developments in the fullness of time.

With respect to the approach itself: one likes to imagine, centuries ago, a mathematician realizing that, contrary to the orthodoxy of the times, there exist functions other than polynomials, and they can be studied, and they are useful. Something not unlike this happened in several European countries in the early thirties; it was realized that some mathematical problems could be better studied by introducing new mathematical entities. Thus, L. C. Young introduced generalized curves for variational problems, Fantappié, and then Schwartz, developed the theory of distributions, of great importance in most areas of mathematics, and DeRham introduced the theory of currents for some variational problems of higher dimensions, in the framework of the theory of differentiable manifolds. In all these developments, the old, classical entities were embedded into the spaces of the new ones, that is, they were considered *as something else* : for instance, L. C. Young considered the usual curves of the calculus of variations as a subset of the set of generalized curves. In this work on control theory, we take the classical elements, the trajectory – control pairs, and inject them into a new space – a space of measures. The theory is developed from there.

This book is inspired by the work of L. C. Young on the calculus of variations (see YOUNG [1], and the references there); as a matter of fact, it could be described as a generalization of his ideas to the rather more demanding problems of optimal control. There have been other attempts at this extension, including the one by L. C. Young himself; ours differs from all these, which are characterized by the definition of a measure on control space together with associated dynamical systems; our approach is based on the definition of a measure on a product space, of which control space is just a factor. In this way, we construct *linear* optimization problems corresponding to control problems normally viewed as nonlinear; and the whole wonderful paraphernalia of linear analysis can be employed to study them!

In the next section we shall review classical control problems, with a view to modifying them.

## 2 Classical control problems

In order to define a classical control problem, we need to describe its several components, such as the differential equation satisfied by the controlled system, the performance criterion, etc. The conditions that we shall put on the functions and sets will serve two purposes. First, they are the kind of reasonable conditions which are usually met when considering classical problems; second, they will allow the modification of these classical problems into others which appear to have some advantages over the classical formulation. To start with, we shall put conditions which permit this with little technical trouble; less restrictive conditions well be introduced later. Let $x$ be a vector in $n$-space $R^n$, $u$ a vector in $m$-space $R^m$, and $t$ a real variable. Consider:

(i) A real closed interval $J = [t_a, t_b]$, with $t_a < t_b$. The interior of this interval in the real line will be denoted by $J^0$, $J^0 = (t_a, t_b)$. This is the time interval in which the controlled system will evolve.
(ii) A bounded, closed, pathwise-connected (see SCHUBERT [1], Part III) set $A$ in $R^n$. The trajectory of the controlled system is constrained to stay in this set for $t \in J$.
(iii) Two elements of $A$, $x_a$ and $x_b$, which are to be the initial and final states of the trajectory of the controlled system.
(iv) A bounded, closed subset $U$ of $R^m$. This is the set in which the control functions are to take values.
(v) Let $\Omega = J \times A \times U$, and $g: \Omega \to R^n$ a continuous function. Consider the differential equation

$$\dot{x}(t) = g[t, x(t), u(t)], \; t \in J^0, \tag{I.1}$$

where the trajectory function $t \in J \to x(t) \in A$ is absolutely continuous and the control function $t \in J \to u(t) \in U$ is Lebesgue-measurable. This is the differential equation describing the controlled system, and is to be satisfied in the sense of Carathéodory; the equality (I.1) holds only Lebesgue-a.e. on $J^0$. (See CODDINGTON AND LEVINSON [1], Chapter 1, for an exposition of these ideas.)
(vi) Let $f_0 : \Omega \to R$ be a continuous function. This function is the integrand of the performance criterion for the problem. Again, we shall be putting other conditions on this function as the work progresses.

We shall say that a trajectory–control pair $p = [x(.), u(.)]$ is *admissible* if the following conditions hold:

(i) The trajectory function $x(.)$ satisfies $x(t) \in A$, $t \in J$, and is absolutely continuous on $J$.
(ii) The control function $u(.)$ takes values in the set $U$, and is Lebesgue-measurable on $J$.

(iii) The boundary conditions $x(t_a) = x_a$, $x(t_b) = x_b$ are satisfied.
(iv) The pair $p$ satisfies the differential equation (I.1) a.e. on $J^0$ in the sense of Carathéodory.

We denote by $W$ the set of admissible pairs. A classical control problem does not have a solution unless this set is nonempty. If this is the case, consider the functional $I$: $W \to R$ defined by

$$I(p) = \int_J f_0[t, x(t), u(t)]\mathrm{d}t, \tag{I.2}$$

where $p = [x(.), u(.)]$. It is desired to minimize this functional over the set $W$. Problems may arise in the quest for the minimizing admissible pair; some of the most important are:

(i) The set $W$ may be empty. It is difficult to determine conditions which guarantee the nonemptiness of the set $W$ when system (I.1) is nonlinear.
(ii) Even if the set $W$ is nonempty, the infimum of $I$ over $W$ may not be achieved at any element of $W$; that is, there may not be a minimizing pair for $I$ in $W$. Conditions which guarantee the existence of a minimum take usually the form of some sort of convexity requirements on sets or functions; these conditions may or may not be artificial when imposed on particular systems. In other words, one may be interested in solving a problem in which the natural descriptions of the system and the functional and sets involved do not satisfy the existence conditions.
(iii) Even if the set $W$ is nonempty, and a minimizing pair for $I$ does exist in $W$, it may be difficult to characterize it; necessary conditions are not always helpful because the information they give may be impossible to interpret.
(iv) The minimizing (or optimal) pair may be very difficult or impossible to estimate numerically; there are no comprehensive computational methods for this purpose.

The situation looks somewhat grim, especially if one is interested in nonstandard problems. Of course linear time-optimal problems and linear problems with quadratic plants can be solved fully. Of course much intelligent effort has been directed towards solving other, nonlinear, problems, sometimes with success. However, one gets the feeling that some of this effort could be better directed towards finding an alternative approach, perhaps one using other spaces, other sets, different things. We shall explore such an approach in the rest of this chapter; in the next section we shall analyse further the classical problem, to gain some inspiration on how to modify it.

## 3 Further analysis of the classical problem

It is necessary at this stage to point out some characteristics of the pairs in $W$, so as to abstract them in due course. Let us first consider the boundary conditions. Let $p = [x(.), u(.)]$ be an admissible pair, and $B$ an open ball in $R^{n+1}$ containing $J \times A$; we denote by $C'(B)$ the space of real-valued continuously differentiable functions on $B$ such that they and their first derivatives are bounded on $B$ (this space is the same as that of all real-valued functions that are uniformly continuous on $B$ together with their first derivatives). Let $\phi \in C'(B)$, and define

$$\phi^g(t, x, u) = \phi_x(t, x)g(t, x, u) + \phi_t(t, x) \tag{I.3}$$

for all $(t, x, u) \in \Omega$; note that both $\phi_x(t, x)$ and $g(t, x, u)$ are $n$-vectors, and that the first term in the right-hand side of (I.3) is their inner product. The function $\phi^g$ is in the space $C(\Omega)$ of real-valued continuous functions defined on the compact product set $\Omega = J \times A \times U$. Since $p = [x(.), u(.)]$ is an admissible pair,

$$\begin{aligned}\int_J \phi^g[t, x(t), u(t)]\mathrm{d}t &= \int_J \{\phi_x[t, x(t)]\dot{x}(t) + \phi_t[t, x(t)]\}\mathrm{d}t \\ &= \int_J \dot{\phi}[t, x(t)]\mathrm{d}t = \phi(t_b, x_b) - \phi(t_a, x_a) \equiv \Delta\phi, \end{aligned} \tag{I.4}$$

for all $\phi \in C'(B)$. It was necessary to introduce the set $B$ and the space $C'(B)$ because $A$ may have an empty interior in $R^n$.

Let us now consider a weak version of (I.1). Denote by $x_j, g_j, j = 1, 2, \ldots, n$, the components of the vector $x$ and function $g$ respectively; let $\mathscr{D}(J^0)$ be the space of infinitely differentiable real-valued functions with compact support in $J^0$ (see TRÈVES [1], Chapter 13, for a treatment of this and related spaces). Define

$$\psi_j(t, x, u) = x_j\psi'(t) + g_j(t, x, u)\psi(t) \tag{I.5}$$

for $j = 1, 2, \ldots, n$, and all $\psi \in \mathscr{D}(J^0)$. Then, if $p = [x(.), u(.)]$ is an admissible pair, we have, for $j = 1, 2, \ldots, n$ and $\psi \in \mathscr{D}(J^0)$,

$$\begin{aligned}\int_J \psi_j[t, x(t), u(t)]\mathrm{d}t &= \int_J x_j(t)\psi'(t)\mathrm{d}t + \int_J g_j[t, x(t), u(t)]\mathrm{d}t \\ &= x_j(t)\psi(t)\Big|_J + \int_J \{\dot{x}_j(t) - g_j[t, x(t), u(t)]\}\psi(t)\mathrm{d}t = 0,\end{aligned} \tag{I.6}$$

since the trajectory and control functions in an admissible pair satisfy (I.1) a.e. on $J^0$, and, since the function $\psi$ has compact support in $J^0$, $\psi(t_a) = \psi(t_b) = 0$.

We note that this equality is not independent of (I.4); it can be derived from it if we choose the function $\phi$ of the form

$$\phi(t, x, u) = x_j\psi(t), (t, x, u) \in \Omega, j = 1, 2, \ldots, n. \quad \text{(I.7)}$$

It is important, however, to single out this special case of the equality (I.4); later on, when considering approximation, we shall be forced to consider problems in which (I.4) is satisfied only for a finite number of functions in $C'(B)$; it will be necessary then to include among these some functions of the type (I.7), so we wish to make sure that we do not overlook these.

The same situation arises for another special choice of functions in $C'(B)$. Put

$$\phi(t, x, u) = \theta(t), (t, x, u) \in \Omega, \quad \text{(I.8)}$$

that is, a function which depends on the time variable only; then $\phi^g(t, x, u) = \dot{\theta}(t)$, $(t, x, u) \in \Omega$, also a function of the time only. We are led thus to consider a subspace of the space $C(\Omega)$, to be denoted by $C_1(\Omega)$, of the functions in this space which depend only on the variable $t$; its elements will be denoted as functions of three variables, $(t, x, u) \to f(t, x, u)$, even if their value does not change when $x$ or $u$, or both, change. Then if $p = [x(.), u(.)]$ is an admissible pair, the equality (I.4) with the choice (I.8) for the function $\phi$ implies that

$$\int_J f[t, x(t), u(t)]\mathrm{d}t = a_f, \quad f \in C_1(\Omega), \quad \text{(I.9)}$$

where $a_f$ is the integral of $f(., x, u)$ over $J$, independent of $x$ and $u$; we have put $f$ for $\dot{\theta}$.

The set of equalities (I.4), of which we have singled out the special cases (I.6) and (I.9), are properties of the admissible pairs in the classical formulation of the optimal control problem; by suitably generalizing them we shall effect the transformation of this into another, nonclassical, problem which appears to have better properties in some respects.

## 4 Metamorphosis

Remember the problems and troubles associated with the classical control problems. The set of admissible pairs may be empty; the functional measuring the performance of the system may not achieve its minimum in this set, if nonempty; it is difficult to obtain information regarding the optimal pair from the necessary conditions; there are no comprehensive ways of estimating numerically the optimal pair. It appears that, at least with respect to the first two difficulties, the situation may become more promising if the set of admissible pairs could somehow be made *larger*; if we could only enlarge this set.

Of course, in a given classical problem, the set of admissible pairs is fixed. If we somehow add elements to it, we are changing the problem, and considering a new one, inspired classically, but a different formulation nevertheless. This is precisely our intention; the basis of this metamorphosis is the fact that an admissible pair $p = [x(.), u(.)]$ can be considered *as something else*, that is, a transformation can be established between the admissible pairs and other mathematical entities; this transformation is an injection, or one–one mapping, so the optimal pair and its image under the transformation can be identified. It is possible then to augment the set of all images of optimal pairs under this transformation. Let $F \in C(\Omega)$, and consider the mapping

$$\Lambda_p \colon F \in C(\Omega) \to \int_J F[t, x(t), u(t)]\mathrm{d}t \in R. \tag{I.10}$$

The following aspects of this mapping are of interest:

(i) It is *well defined*. Since $F$ is continuous on $\Omega$, the integrand in the right-hand side of (I.10) is summable. Note that this mapping is defined by the admissible pair $p$.

(ii) It is *linear*, in the sense that

$$\Lambda_p(\alpha F + \beta G) = \alpha \Lambda_p(F) + \beta \Lambda_p(G),$$

for all $F, G \in C(\Omega)$, all $\alpha, \beta \in R$, a simple consequence of the linearity of the integral. This mapping is, therefore, a linear functional on $C(\Omega)$.

(iii) It is *positive*; that is, if $F(t, x, u) \geqslant 0$ for all $(t, x, u) \in \Omega$, then $\Lambda_p(F) \geqslant 0$.

(iv) It is *continuous* when the space $C(\Omega)$ is given the topology of uniform convergence (see TRÈVES [1], Chapter 11). Continuity here means, then, that there is a constant $K$, independent of the particular function $F$, such that

$$|\Lambda_p(F)| \leqslant K \sup_{\Omega} |F(t, x, u)| \quad \text{for all } F \in C(\Omega).$$

This fact, which can be proved directly with no difficulty, follows actually from (ii) and (iii); a linear positive mapping between the spaces involved must be continuous (see CHOQUET [1], Chapter 8, for a discussion of these matters).

Consider now the transformation $p \to \Lambda_p$, of an admissible pair into a continuous, positive linear functional. We can easily show

**Proposition I.1** The transformation $p \to \Lambda_p$ of the admissible pairs in $W$ into the linear mappings $\Lambda_p$ defined in (I.10) is an injection.

*Proof* We must show that if $p_1 \neq p_2$, then $\Lambda_{p1} \neq \Lambda_{p2}$. Indeed, if $p_j = [x_j(.), u_j(.)]$, $j = 1, 2$, and $p_1 \neq p_2$, then there is a subinterval of $J$, $J_1$ say, on which $x_1(t) \neq x_2(t)$. A continuous function $F$ can be constructed on $\Omega$ so that the right-hand sides of (I.10) corresponding to $p_1$ and $p_2$ are not equal; for instance, make $F$ independent of $u$, equal zero for all $t$ outside $J_1$, and such that it is positive on the appropriate portion of the graph of $x_1(.)$, and zero on that of $x_2(.)$. Then the linear functions are not equal. □

We can identify, therefore, pairs $p$ with linear functionals $\Lambda_p$. It may appear that much has not been achieved with these definitions and discussions, that we are simply writing some integrals in a different way. In reality, we have achieved something deep and useful, when identifying the optimal pairs with positive linear functionals on $\Lambda_p$. Consider the equalities (I.4), (I.6) and (I.9). Their left-hand sides are all integrals of exactly the same type as that appearing in the definition of $\Lambda_p$ in (I.10). These equalities can then be written using the definition of this linear functional:

$$\begin{aligned} \Lambda_p(\phi^g) &= \Delta\phi, \ \phi \in C'(B) \\ \Lambda_p(\psi_j) &= 0, j = 1, 2, \ldots, n, \ \psi \in \mathscr{D}(J^0) \qquad \text{(I.11)} \\ \Lambda_p(f) &= a_f, f \in C_1(\Omega). \end{aligned}$$

Also, the functional $I$ in (I.2) can be written as

$$I(p) = \Lambda_p(f_0). \qquad \text{(I.12)}$$

The image of the set of all admissible pairs $W$ under the transformation $p \to \Lambda_p$ is in the set of all those positive linear functionals on $C(\Omega)$ which satisfy the equalities (I.11). We shall now enlarge this image which, we repeat, can be identified with $W$ itself, and define the new, nonclassical, problem. The classical problem can be rephrased as follows: among those positive linear functionals on $C(\Omega)$ *of the type* $\Lambda_p$ we seek the one for which the number $\Lambda_p(f_0)$ is a minimum. In the new problem, we shall simply consider *all* positive linear functionals $\Lambda$ on $C(\Omega)$ which satisfy (I.11), and seek to minimize the function $\Lambda \to \Lambda(f_0)$ over this new, larger set of functionals. We shall discuss below the reasons for taking this course; before that, we shall introduce measures. In the Appendix we present an introduction to this subject.

Let $\Lambda$ be a linear positive continuous functional on $C(\Omega)$. Some authors (see CHOQUET [1]) define such a functional as a positive Radon measure, the connection with classical measure theory being established later; we shall take this approach. A Radon measure on $\Omega$ can be identified with a regular Borel measure on this set; we treat these matters in some detail in the Appendix, where we prove this result for positive measures and functionals. Thus, given a positive functional $\Lambda$ on $C(\Omega)$ there is a positive Borel measure on $\Omega$ such that

$$\Lambda(F) = \int_\Omega F(t, x, u)\mathrm{d}\mu \equiv \mu(F) \quad \text{for all } F \in C(\Omega);$$

it is said that $\mu$ is a *representing measure* for the functional $\Lambda$. It would be possible to do without the concept of measure in our development, using the functionals themselves. However, it is our belief that some results are visualized better using measures, and measures we shall use. The space of all positive Radon measures on $\Omega$ will be denoted by $\mathscr{M}^+(\Omega)$.

Using these concepts we can put our nonclassical problem in its definitive form, which will be used in the rest of the book. The positive linear functionals above will be replaced by their representing measures; thus, we seek a measure in $\mathscr{M}^+(\Omega)$, to be normally denoted by $\mu^*$, which minimizes the functional

$$\mu \in \mathscr{M}^+(\Omega) \to \mu(f_0) \in R \tag{I.13}$$

subject to the constraints

$$\begin{aligned} \mu(\phi^g) &= \Delta\phi, \ \phi \in C'(B) \\ \mu(\psi_j) &= 0, j = 1, 2, \ldots, n, \ \psi \in \mathscr{D}(J^0) \\ \mu(f) &= a_f, f \in C_1(\Omega). \end{aligned} \tag{I.14}$$

Of course, the choice of this as the nonclassical problem to replace the classical one introduced above begs many questions. Suppose that we can solve it, that we can obtain a positive Radon measure on the set $\Omega$ which satisfies the equalities (I.14) and minimizes (I.13). What do we do with it? After all, we wish to control a plant in an optimal manner by means of a control function while several constraints are satisfied. In what way does the possession of the optimal measure help to achieve this goal, or at least a reasonable modification of it?

In the next and final section of this chapter we shall examine these matters, and indicate how the optimal measure can be used so that a reasonable modification of the original problem comes to be solved; we shall examine the advantages brought about by the new formulation. We shall preview the rest of the book, and explain how we have come, over the last few years, to make these constructions.

## 5 The advantages of the new formulation, and a preview

Let the subset of $\mathscr{M}^+(\Omega)$ defined by the equalities (I.14) be denoted by $Q$. Suppose that an optimal measure $\mu^*$ exists which minimizes the function (I.13) over the set of measures $Q$. We shall prove, after much work in the rest of the book, that there exists a sequence $\{p^j\} = \{[x_R{}^j(.), u^j(.)]\}$ of trajectory–control pairs with the following properties:

(i) As $j \to \infty$, $I(p_j) \to \inf_Q \mu(f_0)$. Since the set $W$ of admissible pairs can be considered, by means of the injection $p \to \Lambda_p$, as a subset of $Q$, we have

$$\inf_{W} I(p) \geqslant \inf_{Q} \mu(f_0); \tag{I.15}$$

we have actually obtained a *global* minimum of the functional $I$ – for sufficiently large values of the index $j$, the pairs in the sequence $\{p^j\}$ give values to the functional $I$ which, if the two greater lower bounds in (I.15) are actually equal, are close to the classical inf $I(p)$ over the set $W$ of admissible pairs. It is possible that this classical infimum is actually larger than the right-hand side of (I.15), the infimum of the functional $\mu \to \mu(f_0)$ over the set of measures $Q$. In such a case, the pairs in the sequence, for sufficiently large values of the index $j$, would perform even better! Better, for instance, than the solution – if one exists – of the classical problem. We do not know under which conditions this situation does arise, if it indeed does arise at all; frankly, we find this problem uninteresting. After all, we are sure that we are doing at least as well with our sequence as in the classical formulation, and we may be doing *considerably better*. Some talk, darkly, of the fact that these two problems – classical and measure-theoretical – may not be *equivalent*, that is, that the expression (I.15) may be a strict inequality. Even better, we say. The opposite view shows that one has much to do, still, to combat orthodoxy.

(ii) These pairs are constructed so that (I.1) is satisfied – that is, $x_R{}^j(.)$ is the response of the system to $u^j(.)$; also, we choose $x(t_a) = x_a$. However, they may not be admissible. It may happen that $x_R{}^j(t_b) \neq x_b$, that is, that the final boundary condition is not satisfied; also, that a portion of the trajectory strays out of the constraint set $A$. (If $x_R{}^j(t_a) = x_a \in A$ is an interior point of $A$, this can only happen away from the initial point, by continuity.) The whole scheme would be quite useless, were it not for the fact that, as the index $j$ tends to $\infty$, *the pairs in the sequence tend to satisfy these constraints*; that is,

$$\lim_{j\to\infty} \|x_R{}^j(t_b) - x_b\| = 0 \quad \lim_{j\to\infty} d[x_R{}^j(t), A] = 0, \ t \in J. \tag{I.16}$$

Here the norm is euclidean in $R^n$, and $d(x, A)$ is the distance between the point $x \in R^n$ and the set $A \subset R^n$.

By considering these measure-theoretical nonclassical problems, it is possible, therefore, to solve a modified type of optimal control problem. Let a controlled system be described by the differential equation (I.1); a closed bounded set $A$ in $R^n$ and a closed bounded set $U$ in $R^m$ are given, together with an initial point $x_a \in A$, and a performance criterion (I.2). A sequence of trajectory–control pairs is sought, $\{p^j\} = \{[x_R{}^j(.), u^j(.)]\}$, so that the trajectory functions are absolutely continuous, the control functions are measurable and take values in the set $U$, the differential equation is satisfied on $J$ in the sense of Carathéodory, $x_R{}^j(t_a) = x_a$, and the limiting conditions

(I.16) are satisfied; as the index $j \to \infty$, the value given to the performance criterion (I.2) should tend to $\inf_Q \mu(f_0)$. This value may be the same as the $\inf_W I(p)$, or it may be less than it.

This is, then, the reasonable modification of the classical problem which will be solved. Whether it is actually reasonable will depend on the kind of application the whole effort is directed to; it will not be reasonable if exact compliance is required with the terminal condition and with the constraint that the trajectory is to stay always in the set $A$. If any deviation at all is allowed in these constraints, then the measure-theoretical method can be used: it is sufficient to choose a member of the optimal sequence $\{p^j\}$ with sufficiently high index $j$. It is of interest to mention that many computational methods developed in a classical framework can only provide estimates of solutions which do not satisfy exactly many of the constraints of the problem; one such is the well known penalty method (see DI PILLO *et al.* [1]). Such methods suffer, in our view, from being associated with a classical, rigid, framework; in contrast, we admit the possibility of the relaxation of the two constraints from the start, and include this possibility in the development of the theory. In this manner, we gain advantages. There are two characteristics of the nonclasical measure-theoretical problems associated with our approach which should be noted:

(i) The existence of an optimal measure in the set $Q$ minimizing $\mu \to \mu(f_0)$ can be studied in a straightforward manner without having to impose conditions such as convexity which may be artificial. We shall consider these problems in Chapter 2, and find interesting relationships between a particular topology on the set $Q$ and existence properties.

(ii) The function $\mu \to \mu(f_0)$, as well as the functions appearing in the left-hand side of the equalities (I.14) – those that define the set $Q \subset \mathcal{M}^+(\Omega)$ – *are linear in their argument, the measure* $\mu$. This fact forms the basis of our approach; since the functions involved are linear even for those problems normally classed as nonlinear, the whole machinery of linear analysis can be used to attack problems such as the development of computational methods and controllability.

In Chapter 3 the basic connection with linear problems will be established; it will be shown that the nonclassical measure-theoretical optimization problem is actually an *infinite-dimensional linear programming problem*, whose solution can be approximated by that of a finite-dimensional one. The basic properties of these solutions are studied in Chapter 4, on approximation, where we shall establish the basic constructions for the optimal sequence $\{p^j\}$ introduced above. Somewhat more restrictive conditions will be put on some of the functions appearing in the problems.

The results of Chapters 3 and 4 are used in Chapter 5 to develop a computational method for the estimation of optimal controls and trajectories. Results are given for several control problems, with up to two state variables and two controls. An interesting consequence of this development has been a *global* method for the minimization of functions of a finite number of variables; examples are given of the application of this method.

The subject of controllability is treated in Chapter 6, where we shall be able to use some very basic and powerful facts from linear analysis for the solution of nonlinear controllability problems.

It is possible to extend most of our results to problems in which the state spaces are Hilbert spaces; these matters are treated in Chapter 7.

In Chapter 8 we study the diffusion equation by means of the same tools used in the rest of the book to treat finite-dimensional systems; we do not make use of the results associated with Hilbert spaces and presented in Chapter 7. A novel approach to the controllability of the diffusion equation, to study which we must go beyond measures, indeed beyond distributions, is an important part of this chapter.

In the Appendix we present background material, especially on the connection between functionals and measures, and on convexity, mainly on the structure of the extremal points of some sets of measures.

It has not been possible to derive a set of useful necessary conditions for the measure-theoretical problem. The fact that we are dealing with a linear programming problem is in this context a hindrance, not a help. We had hoped to characterize somehow the support of the optimal measure; it has proved an elusive undertaking.

Our original approach to these matters, which appeared in several papers in the seventies (RUBIO [1]–4]), was heavily reliant on the characterization of the extremal points of the set of measures $Q$ (the definition of this set was not quite the same as the one we are using now). When we tried to extend this work to problems whose state spaces were Hilbert, we found that some approximation theorems could not be readily extended; it was possible to find a different approach – in which the extremal points played a rather subdued role – to obtain results which were just as interesting; eventually we adopted this approach even for the finite-dimensional case. Better approximation procedures were developed in RUBIO [5], and this work, suitably expanded in subsequent papers, forms the basis of the present book. Most of the relevant material appearing in those early papers has been incorporated in some form in this book.

**References**

General approach: RUBIO [1], [4], [5].
Functional analysis: CHOQUET [1], TRÈVES [1], FRIEDMAN [1].
Measure theory: CHOQUET [1], FRIEDMAN [1].
Distributions: SCHWARTZ [1].
Currents: DE RHAM [1].

# 2
# Existence

## 1 Existence, and a topology

It is a simple matter to prove the existence of an optimal measure in the set $Q$ for the function $\mu \to \mu(f_0)$ under the conditions on the functions and sets of the problem given in Chapter 1; we shall do so in this section. In the next section of this chapter we shall prove an existence theorem under weakened conditions on the control set $U$, which will be allowed to be unbounded.

Existence theorems have a heavy topological content. We first review the main topological ideas, and then apply the main results to our problem. Let $S$ be a subset of a Hausdorff (or separated) topological space $X$, and $y$ a real-valued function on $S$, $y$: $S \to R$. We wish to find out whether there is an element $s^*$ in $S$ such that $y(s^*) \leqslant y(s)$, for all $s \in S$. We shall present a sufficient condition for the existence of such an element, a condition which, we note, is not necessary, but is sufficiently general for our purposes. We make some definitions. (See BERGE [1], SCHUBERT [1], CHOQUET [2].)

(i) We say that the set $S$ is *compact* if it has the *finite intersection property*; that is, let $\{F_\alpha\}$ be any collection of closed sets in $X$ such that any finite number of them has a nonempty intersection; then the total intersection $\cap_\alpha F_\alpha$ is nonempty. We remark that if $X = R^n$, a closed bounded subset of this space is compact.

(ii) We say that the function $y$: $S \to R$ is *lower semicontinuous* (lsc) at a point $s_0$ in $S$ if for each $\varepsilon > 0$ there is a neighbourhood $U_\varepsilon(s_0)$ *in* $S$ such that $y(s) > y(s_0) - \varepsilon$ for all $s \in U_\varepsilon(s_0)$. We say that the function $y$ is *lower semicontinuous in* $S$ if it is lower semicontinuous at every point of $S$. For instance, if $S = [0, 1]$ and $y(s) = 1$ for $y \in (\frac{1}{2}, 1]$, $y(s) = 0$ otherwise, then $y$ is lsc at $s = \frac{1}{2}$, and indeed at every point of $S$; it is then lsc in $S$. If $y(s) = 1$ for $y \in [\frac{1}{2}, 1]$, $y(s) = 0$ otherwise, then $y$ is not lsc at $s = \frac{1}{2}$.

This example shows that not every lsc function is continuous. A continuous function is, however, lsc; if $y$ is continuous at $s_0$, with every $\varepsilon > 0$ we can associate a neighbourhood $U_\varepsilon(s_0)$ in $S$ such that

$|y(s) - y(s_0)| < \varepsilon$ for $s \in U_\varepsilon(s_0)$, which implies the condition for lower semicontinuity.

The relationship between these two concepts, compactness and lower semicontinuity, is immediate:

**Proposition II.1** If $S$ is a compact subset of the Hausdorff space X and the function $y$: $S \to R$ is lsc in $S$, then:

(i) $\inf_S y(s) > -\infty$.

(ii) There is an element $s_0 \in S$ such that $y(s_0) \leqslant y(s)$, for all $s \in S$; that is, the infimum of $y$ is attained on $S$.

*Proof* (i) We show that there is a number $\sigma$ such that $y(s) > \sigma$ for all $s \in S$. Suppose otherwise; then for every positive integer $j$ there is an element $s_j \in S$ such that $y(s_j) < -j$. Consider the sequence $\{s_j\}$; since its elements are in $S$, a compact set, it has at least one adherent point (see CHOQUET [2], Chapter 1); this point, $z$ say, is in $S$. There is a subsequence $\{s_k\}$ converging to $z$. Since the function $y$ is lsc at the point $z$, given $\varepsilon > 0$, for sufficiently large values of the index $k$, $y(s_k) > y(z) - \varepsilon$, which implies $-k > y(z) - \varepsilon$, a contradiction since the value $y(z)$ is a (finite) real number. Thus a number such as $\sigma$ does exist, which implies $\inf_S y(s) > -\infty$.

(ii) Let $\inf_S y(s) \equiv \eta$, and $\lambda$ be a real number such that $\lambda > \eta$. Define

$$V(\lambda) = \{s: s \in S, y(s) < \lambda\}.$$

Then: (a) The set $V(\lambda)$ is nonempty; this is a consequence of the definition of the greatest lower bound (inf) of a set of numbers.

(b) This set is closed. Indeed, let $s_0$ be a point of adherence of $V(\lambda)$, that is, a point in the set $S$ such that every neighbourhood in $S$ of this point intersects $V(\lambda)$. Since $y$ is lsc at $s_0$, for any $\varepsilon > 0$ there is a neighbourhood $U_\varepsilon(s_0)$ in $S$ such that $y(s) > y(s_0) - \varepsilon$ for all $s \in U_\varepsilon(s_0)$. Let $s_1$ be in $V(\lambda) \cap U_\varepsilon(s_0)$. Then $y(s_1) > y(s_0) - \varepsilon$, and

$$y(s_0) < y(s_1) + \varepsilon \leqslant \lambda + \varepsilon \quad \text{for all} \quad \varepsilon > 0.$$

This implies that $y(s_0) < \lambda$, and then $s_0 \in V(\lambda)$. The set $V(\lambda)$ is therefore closed.

(c) If $\lambda_1 < \lambda_2$, $V(\lambda_1) \subset V(\lambda_2)$. An arbitrary *finite* collection of sets $\{V(\lambda_i)\}$ has therefore a nonempty intersection.

(d) Thus, since the set $S$ is compact, the intersection of all sets $V(\lambda)$, for all numbers $\lambda > \eta$, is nonempty. Let $s^*$ be in this intersection. Then $y(s^*)$ must equal $\eta$; otherwise, if $y(s^*)$ were larger than this number, the element $s^*$ would not be in some sets $V(\lambda)$, for values of $\lambda$ sufficiently close to the number $\eta$. Thus, the inf $y(s)$ over the compact set $S$ is attained, at the element $s^* \in S$. □

It is possible now to go back to our original problem, to determine under what conditions there is an optimal Radon measure $\mu$ for the function $\mu \to \mu(f_0)$ in the set $Q$ defined by the equalities (I.14), which we reproduce here:

$$\begin{aligned} \mu(\phi^g) &= \Delta\phi, \ \phi \in C'(B) \\ \mu(\psi_j) &= 0, j = 1, 2, \ldots, n, \ \psi \in \mathscr{D}(J^0) \\ \mu(f) &= a_f, f \in C_1(\Omega). \end{aligned} \tag{II.1}$$

We assume that this set of measures $Q$ is nonempty; it may be, for instance, that the set $W$ of admissible pairs is nonempty, maybe because the system is (classically) controllable. Of course the set $Q$ may be nonempty while $W$ is empty; one of the advantages of the present formulation. We shall study these matters in Chapter 6.

We note, to start with, that no topology has as yet been put on the set $Q$, or on the (linear) space $\mathscr{M}(\Omega)$ of all Radon measures, positive or otherwise, on the set $\Omega$; this space will take on the role of the space $X$ in our results above. We can – and this is a key remark, of importance to the treatment of existence problems in a variety of optimization problems – try to find a Hausdorff topology on this space so that $Q$ is compact and the function $\mu \to \mu(f_0)$ is lsc. Of course, if no optimal measure does exist under our hypothesis, we will never find such a topology; there is an element of risk in this undertaking. As we shall see below, a Hausdorff topology can be found in which the set $Q$ is compact and the function $\mu \to \mu(f_0)$ is not only lsc, but actually continuous. (We shall need the concepts and results connected with lower semicontinuity in Chapter 7; the version of Proposition II.1. in which the function $y$ is assumed continuous is of course simpler to prove.) In this manner we shall prove existence.

There are several ways of characterizing the topology we have in mind, known as the *weak* topology*, or the *vague topology* on the space $\mathscr{M}(\Omega)$. To start with, we note that this space is a linear space, which will become a locally convex topological vector space when given the weak* topology; this can be defined by the family of seminorms $\mu \to |\mu(F)|$, $F \in C(\Omega)$, which gives rise to a basis of neighbourhoods of zero of $\mathscr{M}(\Omega)$:

$$U_\varepsilon = \{\mu \in \mathscr{M}(\Omega) \colon |\mu(F_j)| < \varepsilon, j = 1, 2, \ldots, r\}, \tag{II.2}$$

for every $\varepsilon > 0$ and all finite subsets $\{F_j, j = 1, 2, \ldots, r\}$, $F_j \in C(\Omega)$. Many properties of this topology can be found in the literature (see TRÈVES [1], Chapter 19, CHOQUET [1]). We shall mention only that this is a topology of pointwise convergence; that is, a sequence of measures in $\mathscr{M}(\Omega)$, $\{\mu_i\}$, converges to the zero measure if the numbers $\{\mu_i(F)\}$ converge to zero in the real line, for all $F \in C(\Omega)$. We can prove

**Proposition II.2** (i) The set of measures $Q$, defined as those measures in $\mathscr{M}^+(\Omega)$ which satisfy the equalities (II.1), is compact in the topology induced by the weak* topology on $\mathcal{M}^+(\Omega)$. (ii) The function $\mu \to \mu(f_0)$, mapping $Q$ into the real line, is continuous.

*Proof* (i) It is well known that the set $\{\mu: \mu \in \mathcal{M}^+(\Omega), \mu(1) = \alpha\}$ is compact, since $\Omega$ is compact; here the function 1 equals 1 on $\Omega$, and $\alpha$ is any positive real number (see CHOQUET [1], Theorem 12.6). Amongst the equalities (II.1), we find $\mu(1) = t_b - t_a \equiv \Delta t$; thus, the set $Q$ is a subset of the compact set $\{\mu: \mu \in \mathcal{M}^+(\Omega), \mu(1) = \Delta t\}$. We shall prove that $Q$ is closed, from which the contention that it is compact follows readily. Indeed, $Q$ can be written as an intersection of closed sets:

$$Q = \bigcap_{\phi} \{\mu: \mu(\phi) = \Delta\phi\},$$

remember that the last two sets of equalities in (II.1) are implied by the first set. The set $\{\mu: \mu(\phi) = \Delta\phi\}$ is the inverse image of the (closed) singleton set $\{\Delta\phi\} \subset R$ by the function $\mu \to \mu(\phi)$, which will be proved below to be continuous; this set is thus closed.

(ii) Consider the function $\mu \in Q \to \mu(F) \in R$, for $F \in C(\Omega)$. It is continuous if the inverse image of every neighbourhood of a basis of neighbourhoods of $\mu(F)$ in $R$ is a neighbourhood of $\mu$ in $Q$; that is, if the set $\{\nu \in Q: |(\mu - \nu)(F)| < \varepsilon\}$ is, for all $\varepsilon > 0$, a neighbourhood of $\mu$ in $Q$. This follows from the fact that the weak* topology in $\mathcal{M}(\Omega)$ is defined by the neighbourhoods of zero (II.2), and that the neighbourhoods of a point with respect to the set $Q$ are the intersection of this set with the neighbourhoods of the point in $\mathcal{M}(\Omega)$. □

From Propositions II.1 and II.2 follows our first existence theorem:

**Theorem II.1** There exists an optimal measure $\mu^*$ in the set $Q$ of measures in $\mathcal{M}^+(\Omega)$ which satisfy the equalities (II.1), for which $\mu^*(f_0) \leqslant \mu(f_0)$, for all $\mu \in Q$.

In some earlier papers (see RUBIO [1], [4], [5]), these results are related to the extremal points of a compact convex set; the measure $\mu^*$ is an extreme point of the set $Q$, as we prove in the Appendix. Some approximation schemes are also involved in these developments; we shall not pursue these ideas here. In the following section, we prove an existence theorem under less restrictive hypotheses than those postulated in this section.

## 2 The case of the unbounded controls

We prove now a somewhat more difficult existence theorem, for a control problem with unbounded control set $U$; we follow RUBIO [2], [4] and [6]. Given a closed bounded subset $A$ of $R_n$, an interval $J = [t_a, t_b]$, points $x_a$ and $x_b$ in $A$, and an unbounded closed subset $U$ of $R^m$, we consider the set $W_1$ of admissible pairs $p = [x(.), u(.)]$ satisfying the following conditions:

(i) The trajectory function $x(.)$ satisfies $x(t) \in A$, $t \in J$, and is absolutely continuous on $J$.

(ii) The control function $u(.)$ takes values in the unbounded set $U$, is Lebesgue-measurable on $J$, and is bounded:

$$\sup_J \|u(t)\| < \infty. \tag{II.3}$$

The norm is euclidean in $R^m$. Of course, since the set $U$ is unbounded, no number exists that bounds the left-hand side of (II.3) *for all control functions*.

(iii) The boundary conditions $x(t_a) = x_a$, $x(t_b) = x_b$ are satisfied.

(iv) The pair $p$ satisfies the differential equation (I.1) a.e. on $J^0$ in the sense of Carathéodory.

It is desired to minimize the performance criterion (I.2) over the set $W_1$. The same difficulties encountered in the solution of classical control problems when the control set $U$ was bounded appear now, only, we can guess, in a more acute form. It is, as before, sought to replace the classical formulation by some other, perhaps more promising. The further analysis of the classical problem can be carried out as before, and we conclude that the pairs in $W_1$ satisfy the equalities in (I.4), (I.6), (I.9).

We note that the set $\Omega = J \times A \times U$ is unbounded in this formulation; it is not possible to put the topology of uniform convergence on the space $C(\Omega)$, since the (continuous) functions in this space are not necessarily bounded. Before attempting to inject the set of admissible pairs into a set of continuous functionals we must, therefore, define an appropriate topology on $C(\Omega)$. Of the many available we have found it convenient to use the topology of compact convergence (EDWARDS [1]), defined by the seminorms $P_K$,

$$P_K(F) = \sup_K |F(t, x, u)|, \ K \text{ a compact subset of } \Omega, \ F \in C(\Omega).$$

The functional

$$\Lambda_p\colon F \in C(\Omega) \to \int_J F[t, x(t), u(t)]\mathrm{d}t \in R \tag{II.4}$$

is well defined, linear and positive; it is also continuous since for any $F \in C(\Omega)$

$$|\Lambda_p(F)| \leqslant (t_b - t_a) \sup_K |F(t, x, u)|, \tag{II.5}$$

where $K$ is any compact subset of $\Omega$ containing the set

$$J \times A \times \{u \in U\colon u = u(t), \ t \in J\};$$

note that this set is bounded because of our assumption (II.3) above, and that, according to the theory of locally convex topological vector spaces (see

TRÈVES [1], Chapter 7) a map such as $\Lambda_p$ is continuous if and only if there is *one* continuous seminorm on $C(\Omega)$ such that an inequality such as (II.5) is true.

The transformation $p \to \Lambda_p$ is, again, an injection, and the pairs in $W_1$ can be identified with the linear functionals $\Lambda_p$. Further, each linear positive functional $\Lambda$ on $C(\Omega)$ with the topology of compact convergence can be represented by a regular positive Borel measure *with compact support* (EDWARDS [1]),

$$\Lambda(F) = \int_{\Omega} F(t, x, u)\mathrm{d}\mu \equiv \mu(F);$$

as before, we are led to consider the problem of minimizing the function $\mu \to \mu(f_0)$ over the set – also to be denoted by $Q$ – of all positive regular Borel measures with compact support in $\Omega$ which satisfy the equalities

$$\begin{aligned} \mu(\phi^g) &= \Delta\phi,\ \phi \in C'(B) \\ \mu(\psi_j) &= 0, j = 1, 2, \ldots, n,\ \psi \in \mathscr{D}(J^0) \\ \mu(f) &= a_f, f \in C_1(\Omega) \end{aligned} \tag{II.6}$$

The space of regular Borel measures on $\Omega$ with compact support will be topologized by the $\sigma(C(\Omega)', C(\Omega))$, or weak*, topology. We assume, as when considering the previous case, that the set $Q$ is nonempty. Our first step in the study of the existence of an optimal measure in this set will be one of approximation; we shall consider a sequence of problems such as those treated in Section 1 of this chapter, in which $\Omega$ is compact. Define:

$Q_k = \{\mu$: $\mu$ is a regular positive Borel measure on $\Omega$ with support in the set $J \times A \times \{u \in U, \|u\| \leqslant k\}$ satisfying the equalities (II.6)$\}$

The norm is euclidean in $R^m$. The problems in the sequence consist in minimizing $\mu(f_0)$ over $Q_k$, $k = 1, 2, \ldots$. If $Q_k$ is nonempty, there follows from the results of Section 1:

**Proposition II.3** The set $Q_k$ is compact; the continuous function $\mu \to \mu(f_0)$ attains its minimum on $Q_k$.

With respect to the emptiness or otherwise of the sets $Q_k$, we recall that we have assumed the set $Q$ nonempty, and that the measures in this set have compact support, so that every measure $\mu \in Q$ belongs to some set $Q_k$, for some $k$. Since $Q_{k+1} \supset Q_k$, the sets $Q_k$ are nonempty for $k > k'$, say. Without loss of generality, we can assume $k' = 1$.

Let $\mu_k$ be a measure in $Q_k$ for which $\mu_k(f_0) = \inf_{Q_k} \mu(f_0)$; consider the sequence $\{\mu_k, k = 1, 2, \ldots\}$. We prove a basic result:

**Proposition II.4** Let $\inf_Q \mu(f_0) > -\infty$. Then

$$\lim_{k \to \infty} \mu_k(f_0) = \inf_Q \mu(f_0).$$

*Proof* As mentioned above,

$$Q_1 \subset Q_2 \subset \ldots \subset Q_{k-1} \subset Q_k \subset \ldots \subset Q;$$

then

$$-\infty < \inf_Q \mu(f_0) \leqslant \inf_{Q_k} \mu(f_0) = \mu_k(f_0) \leqslant \mu_{k-1}(f_0) = \inf_{Q_{k-1}} \mu(f_0).$$

Therefore, the sequence $\{\mu_k(f_0), k = 1, 2, \ldots\}$ is nonincreasing and bounded below; it converges, to a number $c \geqslant \inf_Q \mu(f_0)$. Suppose that $c > \inf_Q \mu(f_0)$. Then there is $\nu \in Q$ such that

$$c > \nu(f_0) > \inf_Q \mu(f_0).$$

However, $\nu$ is in $Q$, and has therefore compact support; thus, there is an integer $h$ such that $\nu \in Q_h$. Then,

$$\nu(f_0) \geqslant \inf_{Q_h} \mu(f_0) = \mu_h(f_0) \geqslant c,$$

a contradiction. Thus $c = \inf_Q \mu(f_0)$. □

It is possible, therefore, to approximate $\inf_Q \mu(f_0)$ by the action of the measures $\mu_k$ on the function $f_0$. However, this greatest lower bound is not generally attained over the set $Q$, as in the case when the optimal control is impulsive. Some further conditions are needed to guarantee that this value is attained over $Q$.

We assume, then, that the following conditions are satisfied by the functions $f_0$ and $g$:

(i) The function $f_0$ has the property

$$\frac{f_0(t, x, u)}{\|u\|} \to \infty \tag{II.7}$$

as $\|u\|$ tends to infinity for all $(t, x) \in J \times A$.

(ii) The function $f_0$ is nonnegative on $\Omega$,

$$f_0 \geqslant 0 \text{ on } \Omega \tag{II.8}$$

Since the condition (i) implies that $f_0$ is bounded below on $\Omega$, this is a very mild constraint; nothing much changes if we change the function $f_0$ in the definition of the performance criterion by adding to it a constant so that (II.8) is satisfied on $\Omega$.

(iii) There exist positive constants $c_g$ and $R$ such that

$$\|g(t, x, u)\| \leqslant c_g f_0(t, x, u) \tag{II.9}$$

for $(t, x) \in J \times A$ and $u \in U \cap S_R{}^c$, with $S_R$ the closed ball in $R^m$ with centre at the origin and radius $R$, and $S_R{}^c$ its complement in $R^m$.

**Proposition II.5** If the function $f_0$ satisfies (II.7), then $\inf_Q \mu(f_0) > -\infty$.

*Proof* If $f_0$ satisfies (II.7), it is bounded below. Further, if $\mu \in Q$, it satisfies the equalities (II.6); in particular, $\mu(1) = \Delta t$. Thus, $\mu \in Q$ implies

$$\mu(f_0) \geqslant \mu(1) \inf_\Omega [f_0(t, x, u)] = \Delta t \inf_\Omega [f_0(t, x, u)] > -\infty;$$

then $\inf_Q \mu(f_0) > -\infty$. □

According to this proposition we can apply the results of Proposition II.4 if the function $f_0$ satisfies the condition (II.8). Consider the subspace $E(\Omega)$ of $C(\Omega)$ consisting of those functions $F$ in $C(\Omega)$ with the following property:

There exist constants $C > 0$ and $R > 0$, possibly depending on the particular function $F$, such that $|F(t, x, u)| \leqslant Cf_0(t, x, u)$, $(t, x, u) \in J \times A \times U \cap (S_R{}^c)$.

Let $\bar{E}(\Omega)$ be the closure of this subspace in $C(\Omega)$. We put on $\bar{E}(\Omega)$ the topology induced by $C(\Omega)$; then $\bar{E}(\Omega)$, a closed subspace of the Fréchet space $C(\Omega)$ with the induced topology, is a Fréchet space. (A Fréchet space is a complete, metrizable, locally convex topological vector space.) Our main interest here is in the applicability of the Banach–Steinhaus theorem. (See TRÈVES [1], Chapter 10.) We can prove

**Proposition II.6** The functionals

$$F \to \mu_k(F),\ F \in \bar{E}(\Omega),\ k = 1, 2, \ldots$$

are continuous.

*Proof* We recall the definition of the measures $\mu_k$, $k = 1, 2, \ldots$, as those that minimize $\mu(f_0)$ over $Q_k$, $k = 1, 2, \ldots$. Since the functionals $F \to \mu_k(F)$, $F \in C(\Omega)$, $k = 1, 2, \ldots$, are continuous, and the topology on $\bar{E}(\Omega)$ is that induced by $C(\Omega)$, the contention follows. □

**Proposition II.7** The set $Y = \{\mu_k, k = 1, 2, \ldots\}$ is relatively compact in the weak* topology $\sigma(\bar{E}(\Omega)', \bar{E}(\Omega))$.

*Proof* A subset of a topological space is said to be relatively compact if its closure is compact. To prove this proposition we shall use a result which is a consequence of the Banach–Steinhaus theorem; since $\bar{E}(\Omega)$ is a Fréchet space, it is barrelled, and then a subset of its dual $\bar{E}(\Omega)'$ is relatively compact in the weak* topology if and only if it is weakly*–bounded (TRÈVES [1], Chapter 33). We prove that the set $Y$ is weakly*–bounded, and then the proposition follows.

Let $\Omega' = J \times A \times S_R \cap U$, $\Omega'' = J \times A \times S_R{}^c \cap U$. Assume first that $F \in E(\Omega)$. Then,

$$|\mu_k(F)| = \left|\int_\Omega F\mathrm{d}\mu_k\right| \leqslant \int_\Omega |F|\mathrm{d}\mu_k = \int_{\Omega'} |F|\mathrm{d}\mu_k + \int_{\Omega''} |F|\mathrm{d}\mu_k.$$

Also, since $\mu(1) = \Delta t$, and $\mu$ is a positive measure,

$$\int_{\Omega'} |F|\mathrm{d}\mu_k \leqslant \sup_{\Omega'} |F(t, x, u)| \int_{\Omega'} \mathrm{d}\mu_k \leqslant \sup_{\Omega'} |F(t, x, u)| \int_\Omega \mathrm{d}\mu_k$$
$$= \Delta t \sup_{\Omega'} |F(t, x, u)|.$$

Besides, since $F \in E(\Omega), f_0 \geqslant 0$ on $\Omega$ and $\mu_1(f_0) \geqslant \mu_k(f_0)$, $k = 1, 2, \ldots,$

$$\int_{\Omega''} |F|\mathrm{d}\mu_k \leqslant c \int_\Omega f_0\mathrm{d}\mu_k = c\mu_k(f_0) \leqslant c\mu_1(f_0).$$

Thus,

$$|\mu_k(F)| \leqslant \Delta t \sup_{\Omega'} |F(t, x, u)| + c\mu_1(f_0);$$

the set $\{|\mu(F)|, k = 1, 2, \ldots\}$ is bounded for all $F$ in $E(\Omega)$, with the bound depending in general on the particular function $F$.

Since the space $C(\Omega)$ is first countable, it is enough to use sequences, rather than the more general nets, to characterize those elements in $\bar{E}(\Omega)$ not in $E(\Omega)$ itself. Let then $F$ be the limit of a sequence $\{F_i\}$ of functions in $E(\Omega)$. Suppose that $\{|\mu_k(F)|, k = 1, 2, \ldots\}$ is not a bounded set. Then, given any number $R_1$, we can find an integer $h$ such that $|\mu_h(F)| > R_1$. Since $\{F_i\} \to F$ in $C(\Omega)$ and the functionals $G \to \mu_k(G)$, $G \in C(\Omega)$, are continuous, this implies that $|\mu_h(F_j)| > R_1$ for some index $j$, a contradiction.

Thus, for any $F \in E(\Omega)$ the set $\{|\mu_k(F)|, k = 1, 2, \ldots\}$ is bounded, with the bound depending in general on the function $F$. A subset $B$ of a topological vector space (in this case, $\mathscr{M}(\Omega)$ with the weak* topology) is said to be bounded (weakly*-bounded, in this case) if to *every* neighbourhood of zero $V$ in some basis of neighbourhoods a number $\lambda > 0$ can be associated such that $B \subset \lambda V$. A subset $\mathscr{M}_1$ of $\mathscr{M}(\Omega)$ – other than the singleton set $\{0\}$ – is therefore weakly*-bounded if for every set of the form (II.2) a number $\lambda > 0$ can be found such that, for every measure $\mu$ in $\mathscr{M}_1$, $|\mu(F_j)| < \varepsilon/\lambda$; that is, if every number of the form $|\mu(F)|$ is bounded above by a bound independent of the particular measure $\mu \in \mathscr{M}_1$, but which may depend on the function $F \in C(\Omega)$. Thus, the set $Y$ is weakly*-bounded. □

**Proposition II.8** The functions $\phi^g$, $\phi \in C'(B)$, $\psi_j$, $j = 1, 2, \ldots, n$, $\psi \in \mathscr{D}(J^0)$, $f \in C_1(\Omega)$, are in $\bar{E}(\Omega)$.

*Proof* These functions are all in the subspace $E(\Omega)$. Indeed, it follows from the definition of the functions $\phi^g$ in (I.3) that

$$\begin{aligned} |\phi^g(t, x, u)| &\leqslant \|\phi_x(t, x)\| \, \|g(t, x, u)\| + |\phi_t(t, x)| \\ &\leqslant C_\phi \|g(t, x, u)\| + C_\phi \\ &\leqslant c_1 \|g(t, x, u)\|, \end{aligned}$$

for $(t, x, u) \in J \times A \times U \cap S_P{}^c$, for some positive constants $c_1$ and $P$. Here the norm is euclidean, and $C_\phi$ is an upper bound for the norms and absolute values of the derivatives of the function $\phi$ over $J \times A$. Since $g$ satisfies (II.9), we have finally

$$|\phi^g(t, x, u)| \leqslant c_1 c_g f_0(t, x, u)$$

for $(t, x, u) \in J \times A \times U \cap S_T{}^c$, $T = \max(R, P)$. The other functions are, as shown in Chapter 1, special cases of the functions $\phi^g$. □

We can prove the main existence theorem:

**Theorem II.2** Under the conditions (II.7), (II.8) and (II.9) on the functions $f_0$ and $g$, the $\inf_Q \mu(f_0)$ is attained over $Q$.

*Proof* By Proposition II.7, the closure of $Y = \{\mu_k, k = 1, 2, \ldots\}$ in $\bar{E}(\Omega)'$ is weakly*-compact. There exists then a subsequence of $Y$, to be denoted also by $\{\mu_k, k = 1, 2, \ldots\}$, which converges weakly* to a functional $\nu^*$ in $\bar{E}(\Omega)'$. In particular, and using the result of Proposition II.4,

$$\lim_{k \to \infty} \mu_k(f_0) = \nu^*(f_0) = \inf_Q \mu(f_0).$$

We must show, of course, that $\nu^*$ – rather, an extension of this functional – is in $Q$. Indeed, all equalities in (II.6) are satisfied by $\nu^*$, since the measures $\mu_k$ satisfy these equalities, $\mu_k \to \nu^*$ in the $\sigma(\bar{E}(\Omega)', \bar{E}(\Omega))$ topology and the functions $\phi^g$ are in $\bar{E}(\Omega)$, by Proposition II.8. Also, the functional $\nu^*$ is positive. Finally, since it is in $\bar{E}(\Omega)'$ – that is, it is continuous – it can be extended, by the Krein–Rutman theorem, to the whole of $C(\Omega)$ as a positive continuous functional there. (See SCHAEFER [1], Theorem 5.4, Corollary 2.) Indeed, according to this theorem, this extension can be effected if the subspace $\bar{E}(\Omega)$ contains functions which are strictly positive on $\Omega$; this is certainly the case according to the definition of this subspace given above. Let the extension of $\nu^*$ be represented by a regular positive Borel measure with compact support, $\mu^*$. We have shown that $\mu^*$ is in the set $Q$. □

Since the support of the measure $\mu^*$ is compact, it is a set of the form $J \times A \times U_c$, with $U_c$ a compact subset of the control set $U$. Thus, it is sufficient to prove approximation theorems for the case when $U$ is compact, since they automatically apply to the present situation.

The condition (II.9) on the function $g$ required by this theorem for the existence of an optimal measure in the set $Q$ is weaker than that required in CESARI [1]; there, in a different framework, is imposed the condition (in our notation)

$$\|g(t, x, u)\| \leqslant C + D\|u\|,$$

for some nonnegative constants $C$ and $D$, and $(t, x, u) \in \Omega$. This is not surprising, since Cesari's definition of what is accepted as a solution of the optimal control problem is stronger than ours.

Still another existence theorem for these problems is proved in RUBIO [7], but we will present it in Chapter 7, when we treat systems with state spaces which are Hilbert. The value of the hypothesis under which we prove this theorem is more readily appreciated in this more general case; also, the main result is closely related to the problems in approximation which we will cover in Chapter 4.

In the next chapter we shall start the study of these matters, when we consider the measure-theoretical optimization problem as an infinite-dimensional linear programming problem; in this manner, we shall derive our first results on approximation.

## References

Existence theorems: RUBIO [2], [4], [6], [7], CESARI [1].
Topology: BERGE [1], SCHUBERT [1], CHOQUET [2].

# 3
# Linear programs

## 1 Introduction

We shall assume in this chapter that the control set $U$ is compact. Consider again the problem of minimizing the linear function

$$\mu \to \mu(f_0) \qquad \text{(III.1)}$$

over the set $Q$ of positive Radon measures on $\Omega$ which satisfy the equalities

$$\begin{aligned} \mu(\phi^g) &= \Delta\phi, \ \phi \in C'(B) \\ \mu(\psi_j) &= 0, j = 1, 2, \ldots, n, \ \psi \in \mathscr{D}(J^0) \qquad \text{(III.2)} \\ \mu(f) &= a_f, f \in C_1(\Omega). \end{aligned}$$

As noted someplace in Chapter 1, this problem is one of *linear programming*; all the functions in (III.1) and (III.2) are linear in the variable $\mu$, and, furthermore, the measure $\mu$ is required to be positive. We note again that this is so even if the original problem is nonlinear; linearity in the present sense was gained by the consideration of admissible pairs as positive measures on $\Omega$, and by the extension of the set of measures thus generated to a larger set, the set $Q$ described by the equalities (III.2). It could be expected that this achievement – the transformation of a nonlinear control problem into a linear optimization one – would bring advantages, perhaps of a practical kind; that ways would be found for resolving some of the difficulties encountered when considering classical problems.

The linear programming problem consisting in minimizing the function (III.1) on the subset $Q$ of $\mathscr{M}^+(\Omega)$ described by the equalities (III.2) is not finite-dimensional; the underlying space, $\mathscr{M}(\Omega)$, is not finite-dimensional, and the number of equalities in (III.2) is not finite. We say that it is an *infinite-dimensional linear programming problem*. There is a large and growing literature on such problems (see DUFFIN [1], one of the pioneering works, as a good introduction); we shall be mainly interested in approximations, such as those discussed by VERSHIK [1]. It is possible to approximate the solution of this problem by the solution of a finite-dimensional linear program of sufficiently large dimensions. Also, the solution of this new problem can be approximated in a suitable way by means of a trajectory–control pair. This pair may not be strictly admissible, but, as discussed in detail in Chapter 1, it can be chosen so that the conditions of admissibility tend to be satisfied. By increasing the dimensionality of the finite-dimensional linear programming problem, as well as the accuracy of the computation, the degree to which these conditions are satisfied can be increased. The end product will be a *minimizing sequence* $\{p^j, j = 1, 2, \ldots\}$, with all the good properties described in Section 5 of Chapter 1.

In this chapter as well as in Chapter 4, these matters will be discussed in detail and proved.

## 2 A first approximation

We shall first develop an intermediate program, still infinite-dimensional, by considering the minimization of $\mu \to \mu(f_0)$ not over the set $Q$ but over a subset of $\mathscr{M}^+(\Omega)$ defined by requiring that only a finite number of the constraints in (III.2) be satisfied. This will be achieved by choosing countable sets of functions whose linear combinations are dense in the appropriate spaces, and then selecting a finite number of these.

Consider the first set of equalities in (III.2). Let the set $\{\phi_i, i = 1, 2, \ldots\}$ be such that the linear combinations of the functions $\phi_i \in C'(B)$ are uniformly dense – that is, dense in the topology of uniform convergence – in the space $C'(B)$. (Such a set is said to be *total* in $C'(B)$.) For instance, these functions can be taken to be monomials in the components of the $n$-vector $x$ and the variable $t$. The second and third sets of equalities in (II.2) are, as mentioned often before, not independent, but just special cases of the equalities in the first set; it will be necessary in the following development, however, to deal with the functions appearing in these sets in a manner apart from the rest of the functions in $C'(B)$.

Consider the functions in $\mathscr{D}(J^0)$ defined by

$$\sin\,[2\pi r(t - t_a)/\Delta t],\ 1 - \cos\,[2\pi r(t - t_a)/\Delta t],\ r = 1, 2, \ldots, \quad \text{(III.3)}$$

where $\Delta t = t_b - t_a$. We shall call $\{\chi_h, h = 1, 2, \ldots\}$ the sequence of functions of the type $\psi_j(t, x, u) = x_j\psi'(t) + g_j(t, x, u)\psi(t)$, defined in (I.5), when the functions $\psi$ are the sine and cosine functions (III.3) defined above and $j = 1, 2, \ldots, n$. We can prove

**Proposition III.1** Consider the linear program consisting in minimizing the function $\mu \to \mu(f_0)$ over the set $Q(M_1, M_2)$ of measures in $\mathscr{M}^+(\Omega)$ satisfying

$$\begin{aligned} \mu(\phi^g{}_i) &= \Delta\phi_i,\ i = 1, 2, \ldots, M_1, \\ \mu(\chi_h) &= 0,\ h = 1, 2, \ldots, M_2. \end{aligned} \quad \text{(III.4)}$$

As $M_1$ and $M_2$ tend to infinity,

$$\eta(M_1, M_2) \equiv \inf_{Q(M_1, M_2)} \mu(f_0)$$

tends to

$$\eta = \inf_{Q} \mu(f_0).$$

*Proof* (i) We prove first that the sequence $\{\eta(M_1, M_2), M_1 = 1, 2, \ldots, M_2 = 1, 2, \ldots\}$ converges as $M_1$, $M_2$ tend to infinity. Consider first the sequence $\{\eta(M_1, M_1), M_1 = 1, 2, \ldots\}$. Since

$$Q(1, 1) \supset Q(2, 2) \supset \ldots \supset Q(M_1, M_1) \supset \ldots \supset Q,$$

$$\eta(1, 1) \leqslant \eta(2, 2) \leqslant \ldots \leqslant \eta(M_1, M_1) \leqslant \ldots \leqslant \eta,$$

this sequence is nondecreasing and bounded above; it converges to a number $\xi \leqslant \eta$. Let $\varepsilon > 0$. For $M_1 > N(\varepsilon)$,

$$|\eta(M_1, M_1) - \xi| < \varepsilon. \tag{III.5}$$

Consider now $\eta(M_1, M_2)$, for both $M_1$ and $M_2$ larger than $N(\varepsilon)$. Without loss of generality, assume that $M_1 > M_2$. Then,

$$\eta(M_2, M_2) \leqslant \eta(M_1, M_2) \leqslant \eta(M_1, M_1),$$

that is,

$$\eta(M_2, M_2) - \xi \leqslant \eta(M_1, M_2) - \xi \leqslant \eta(M_1, M_1) - \xi,$$

or

$$|\eta(M_1, M_2) - \xi| \leqslant \varepsilon.$$

Thus, the double sequence $\{\eta(M_1, M_2)\}$ does converge as $M_1$ and $M_2$ tend to infinity, to the number $\xi$.

(ii) We must prove now that this limit $\xi$ equals $\eta$, the $\inf_Q \mu(f_0)$. We first show that the double limit $\xi$ can be computed sequentially. It is known that

$$\xi = \lim_{M_1 \to \infty} \left[ \lim_{M_2 \to \infty} \eta(M_1, M_2) \right],$$

provided that the $\lim_{M_2 \to \infty} \eta(M_1, M_2)$ exists. To show the existence of this limit, we fix $M_1$ and vary $M_2$; then,

$$Q(M_1, 1) \supset Q(M_1, 2) \supset \ldots \supset Q(M_1, M_2) \supset \ldots,$$

$$\eta(M_1, 1) \leqslant \eta(M_1, 2) \leqslant \ldots \leqslant \eta(M_1, M_2) \leqslant \ldots,$$

for $M_1 = 1, 2, \ldots$. The sequence $\{\eta(M_1, M_2), M_2 = 1, 2, \ldots\}$ is nondecreasing and bounded above; it converges, to a number $\zeta(M_1)$. The double limit $\xi$ *can* be computed sequentially; this is useful.

(iii) Define now

$$Q(M_1) = \bigcap_{M_2=1}^{\infty} Q(M_1, M_2);$$

then

$$\zeta(M_1) = \lim_{M_2 \to \infty} \eta(M_1, M_2) = \inf_{Q(M_1)} \mu(f_0).$$

Also,

$$Q(1) \supset Q(2) \supset \ldots \supset Q(M_1) \supset \ldots \supset Q,$$

that is,

$$\zeta(1) \leqslant \zeta(2) \leqslant \ldots \leqslant \zeta(M_1) \leqslant \ldots \leqslant \eta;$$

as expected, the sequence $\{\zeta(M_1)\}$ converges, necessarily to the same number $\xi$ introduced above in (III.5), and $\xi \leqslant \eta$.

(iv) We can now prove that $\xi = \eta$. Let

$$P = \bigcap_{M_2=1}^{\infty} Q(M_1).$$

Then $P \supset Q$, and

$$\xi = \lim_{M_1 \to \infty} \zeta(M_1) = \inf_P \mu(f_0).$$

We can show that under the conditions of the problem $Q \supset P$, that is, $Q = P$, which will finally imply that

$$\zeta = \lim_{M_1 \to \infty} \zeta(M_1) = \inf_Q \mu(f_0),$$

which is the contention in the theorem, since $\xi$ is the limit of the double sequence $\{\eta(M_1, M_2)\}$.

Indeed, we prove that if $\mu \in P$, then it is also in $Q$, from which the contention, and the proposition, follow. If $\mu \in P$, $\mu(\phi) = \Delta\phi$ for all functions $\phi$ in the subspace spanned by the set $\{\phi_i, i = 1, 2, \ldots\}$, by linearity. This implies that this equality holds for all $\phi \in C'(B)$, since for such functions $\phi$ there exists a sequence $\{\phi^j\}$ of functions in this subspace for which

$$\sup_B |\phi_t(t, x) - \phi_t^j(t, x)|, \sup_B \| \phi_x(t, x) - \phi_x^j(t, x) \|, \sup_B |\phi(t, x) - \phi^j(t, x)|$$

tend to zero as $j$ tends to infinity. Then,

$$\begin{aligned} |\mu(\phi^g) - \Delta\phi| &= |\mu(\phi^g) - \Delta\phi - \mu(\phi^{jg}) + \Delta\phi^j| \\ &= \bigg| \int_\Omega \{[\phi_x(t, x) - \phi_x^j(t, x)]g(t, x, u) \\ &\quad + [\phi_t(t, x) - \phi_t^j(t, x)]\}\mathrm{d}\mu - (\Delta\phi - \Delta\phi^j) \bigg| \\ &\leqslant K_1 \sup_B \| \phi_x(t, x) - \phi_x^j(t, x) \| \\ &\quad + K_2 \sup_B |\phi_t(t, x) - \phi_t^j(t, x)| + K_3 \sup_B |\phi(t, x) - \phi^j(t, x)|; \end{aligned}$$

then $\mu(\phi) = \Delta\phi$, since the last in this chain of expressions tends to zero as $j$ tends to infinity, while the first is independent of this index.

If $\mu \in P$, $\mu(\chi) = 0$ for all functions $\chi$ in the subspace spanned by the set $\{\chi_h, h = 1, 2, \ldots\}$ appearing in (III.4). This implies that $\mu(\psi_j) = 0$ for all $\psi \in \mathscr{D}(J^0)$ and $j = 1, 2, \ldots, n$, since: (a) If $\psi \in \mathscr{D}(J^0)$, its Fourier series converges uniformly on any subinterval in $J$; (b) the Fourier series for $\psi'$ also converges uniformly on any subinterval of $J$ (see CARSLAW [1]); (c) any function $\psi_j$ can be approximated uniformly on $\Omega$, therefore, by a sequence of functions in the subspace spanned by the set $\{\chi_h\}$; (d) the rest of the proof is much like that in the previous case.

We have proved that $P = Q$, and then the contention of the proposition, on the limit of that double sequence, follows. □

## 3 Finite dimensions

The first stage of the approximation scheme has been successfully completed; this is the beginning, one of several such stages. We have limited the number of constraints in the original linear program; the underlying space is not, however, finite-dimensional. It is possible, though, to develop a finite-dimensional linear program whose solution can be used to construct one for the problem of minimizing $\mu \to \mu(f_0)$ over the set (III.4); this solution happens to have a very simple structure.

It will be assumed that the first function appearing in the set of equalities (III.4) is $\phi^g{}_1 = 1$ for $(t, x, u) \in \Omega$; as before, the first equality will be written $\mu(1) = \Delta t$. We write $z$ for the triple $(t, x, u) \in \Omega$. A unitary atomic measure with support the singleton set $\{z\}$, to be denoted by $\delta(z) \in \mathscr{M}^+(\Omega)$, is characterized by

$$\delta(z)(F) = F(z), F \in C(\Omega), z \in \Omega.$$

It is possible to characterize a measure in the set $Q(M_1, M_2)$ at which the linear function $\mu \to \mu(f_0)$ attains its minimum; it follows from a result of ROSENBLOOM [1], which we prove in Theorem A.5 of the Appendix, that:

**Proposition III.2** The measure $\mu^*$ in the set $Q(M_1, M_2)$ at which the function $\mu \to \mu(f_0)$ attains its minimum has the form

$$\mu = \sum_{k=1}^{M_1+M_2} \alpha_k^* \delta(z_k^*), \qquad \text{(III.6)}$$

with the triples $z_k^* \in \Omega$, and the coefficients $\alpha_k^* \geqslant 0, k = 1, 2, \ldots, M_1 + M_2$.

This structural result points the way towards a further approximation scheme; the measure-theoretical optimization problem is equivalent to a nonlinear optimization problem, in which the unknowns are the coefficients $\alpha_k^*$ and supports $\{z_k^*\}$, $k = 1, 2, \ldots, M_1 + M_2$. We have examined this option in earlier papers (RUBIO [1], [4]; however, we shall take a different road this time, and try somehow to preserve the essential linearity of the problem. It would be convenient if we could minimize the function $\mu \to \mu(f_0)$ only with respect to the coefficients $\alpha_k$ in (III.6) – this would be a linear programming problem. However, we do not know the support of the optimal measure. The answer lies in approximating this support, by introducing a set dense in $\Omega$:

**Proposition III.3** Let $\omega$ be a countable dense subset of $\Omega$. Given $\varepsilon > 0$, a measure $\nu \in \mathscr{M}^+(\Omega)$ can be found such that

$$|(\mu^* - \nu)f_0| < \varepsilon, |(\mu^* - \nu)\phi^g{}_i| < \varepsilon, i = 1, \ldots, M_1,$$
$$|(\mu^* - \nu)\chi_h| < \varepsilon, h = 1, \ldots, M_2; \qquad \text{(III.7)}$$

the measure $\nu$ has the form

$$\nu = \sum_{k=1}^{M_1+M_2} \alpha_k^* \delta(z^k),$$

where the coefficients $\alpha_k^*$ are the same as in the optimal measure (III.6), and $z^k \in \omega, k = 1, 2, \ldots, M_1 + M_2$.

*Proof* Consider a countable set $\omega$, dense in $\Omega$. Given $\varepsilon > 0$, we can find

$$z^k \in \omega, \ k = 1, 2, \ldots, M_1 + M_2,$$

such that the inequalities (III.7) are satisfied; indeed, write $f_i = \phi^g{}_i$, $i = 1, 2, \ldots, M_1, f_i = \chi_h, i = M_1 + h, h = 1, 2, \ldots, M_2$. Then, for $i = 0, 1, 2, \ldots, M_1 + M_2$,

$$|(\mu^* - \nu)f_i| = \left| \sum_{k=1}^{M_1+M_2} \alpha_k^*[f_1(z_k^*) - f_i(z^k)] \right|$$
$$\leqslant \Delta t \max_{i,k} |f_i(z_k^*) - f_i(z^k)|;$$

the $\max_{i,k}$ can be made less than $\varepsilon/\Delta t$ by choosing $z^k, k = 1, 2, \ldots, M_1 + M_2$, sufficiently near to $z_k^*$; note that the functions $f_i$, $i = 0, 1, 2, \ldots, M_1 + M_2$, are continuous, and that there is a finite number of them only. The inequalities (III.7) follow. □

This result suggests that the following linear program should be considered. Given $\varepsilon > 0$ and $z_j, j = 1, 2, \ldots, N, z_j \in \omega$, where $\omega$ is a dense subset of $\Omega$, minimize

$$\sum_{j=1}^{N} \alpha_j f_0(z_j) \qquad \text{(III.8)}$$

on the set $P(M_1, M_2)^\varepsilon$ in $R^N$ defined by

$$\alpha_j \geqslant 0, j = 1, 2, \ldots, N,$$

$$-\varepsilon \leqslant \sum_{j=1}^{N} \alpha_j \phi^g{}_i(z_j) - \Delta\phi_i \leqslant \varepsilon, i = 1, 2, \ldots, M_1, \qquad \text{(IIII.9)}$$

$$-\varepsilon \leqslant \sum_{j=1}^{N} \alpha_j \chi_h(z_j) \leqslant \varepsilon, h = 1, 2, \ldots, M_2.$$

Note that the elements $z_j, j = 1, 2, \ldots, N$, are fixed; the only unknowns are the numbers $\alpha_j, j = 1, 2, \ldots, N$. There are $N$ unknowns and $2(M_1 + M_2)$ inequalities in this linear programming problem.

The argument presented in the proof of Proposition III.3 indicated – gave the hope – that the problem of minimizing $\mu \to \mu(f_0)$ over the set $Q(M_1, M_2)$ defined by the equalities (III.4) is closely related to that of minimizing the linear function (III.8) over the set $P(M_1, M_2)^\varepsilon$ defined by the inequalities (III.9), at least for sufficiently large $N$ and small $\varepsilon$. Indeed, our main result in this chapter shows that this is the case.

**Theorem III.1** For every $\varepsilon > 0$, the problem of minimizing the functional (III.8) on the set $P(M_1, M_2)^\varepsilon$ described by the inequalities (III.9) has a solution for $N = N(\varepsilon)$ sufficiently large. The solution satisfies

$$\eta(M_1, M_2) + \rho(\varepsilon) \leqslant \sum_{j=1}^{N} \alpha_j f_0(z_j) \leqslant \eta(M_1, M_2) + \varepsilon, \qquad \text{(III.10)}$$

where $\rho(\varepsilon)$ tends to zero as $\varepsilon$ tends to zero.

*Proof* (i) Given $\varepsilon > 0$, a set

$$\{z^k, k = 1, 2, \ldots, M_1 + M_2\} \qquad \text{(III.11)}$$

can be found, as in the proof of Proposition III.3, so that the inequalities (III.7) are satisfied. For $N$ sufficiently large, the set

$$\omega_N \equiv \{z_j, j = 1, 2, \ldots, N\} \subset \omega$$

will contain (III.11); thus, the set $P(M_1, M_2)^\varepsilon$ is nonempty for such values of $N$, since the $N$-tuple

$$\{\alpha_k{}^*, k = 1, \ldots, M_1 + M_2, 0, \ldots, 0\}$$

is in this set. Since the first inequality in (III.9) is

$$-\varepsilon \leqslant \sum_{j=1}^{N} \alpha_j - \Delta t \leqslant \varepsilon,$$

this set of $N$-tuples with nonnegative entries is bounded; it is also closed, and thus compact. The linear function (III.8) attains its minimum over this set, and

$$\min \sum_{j=1}^{N} \alpha_j f_0(z_j) \leqslant \sum_{k=1}^{M_1+M_2} \alpha_k f_0(z^k) \leqslant \eta(M_1, M_2) + \varepsilon; \qquad \text{(III.12)}$$

one of the inequalities (III.10) has been proved. The other is harder. Let

$$Q(M_1, M_2)^\varepsilon = \{\mu \colon \mu \in M^+(\Omega), |\mu(\phi^g{}_i) - \Delta\phi_i| \leqslant \varepsilon, i = 1, 2, \ldots, M_1, \\ |\mu(\chi_h)| \leqslant \varepsilon, h = 1, 2, \ldots, M_2\}.$$

Then, the set of measures of the type

$$\mu = \sum_{j=1}^{N} \alpha_j \delta(z_j),$$

with the coefficients $\alpha_j$ in the set $P(M_1, M_2)^\varepsilon$, is a subset of $Q(M_1, M_2)^\varepsilon$. Thus,

$$\min \sum_{j=1}^{N} \alpha_j f_0\,(z_j) \geqslant \min \mu(f_0), \qquad \text{(III.13)}$$

where the minimum in the left-hand side of this inequality is over the set $P(M_1, M_2)^\varepsilon$, and that in the right-hand side over $Q(M_1, M_2)^\varepsilon$.
Also,

$$Q(M_1, M_2) = \bigcap_{\varepsilon>0} Q(M_1, M_2)^\varepsilon,$$

and

$$Q(M_1, M_2)^{\varepsilon_1} \supset Q(M_1, M_2)^{\varepsilon_2}, \text{ if } \varepsilon_1 > \varepsilon_2.$$

Let $\eta(M_1, M_2, \varepsilon)$ equal the infimum of $\mu(f_0)$ over the set of measures $Q(M_1, M_2)^\varepsilon$. Then,

$$\eta(M_1, M_2, \varepsilon_1) \leqslant \eta(M_1, M_2, \varepsilon_2), \text{ if } \varepsilon_1 > \varepsilon_2$$

It is sufficient for our purposes to consider a sequence of values of $\varepsilon$, $\varepsilon = 1/p$, $p = 1, 2, \ldots$. Then,

$$\eta(M_1, M_2, 1/1) \leqslant \eta(M_1, M_2, 1/2) \leqslant \ldots \leqslant \eta(M_1, M_2, 1/p) \leqslant \ldots;$$

the sequence $\{\eta(M_1, M_2, 1/p)\}$ is nondecreasing and bounded above. It converges, therefore, to a number $\gamma(M_1, M_2)$ satisfying

$$\gamma(M_1, M_2) = \lim_{p\to\infty} \eta(M_1, M_2, 1/p) = \inf_{Q(M_1, M_2)} \mu(f_0) = \eta(M_1, M_2).$$

Thus,

$$\rho(\varepsilon) \equiv \eta(M_1, M_2, \varepsilon) - \eta(M_1, M_2) \qquad \text{(III.14)}$$

tends to zero as $\varepsilon$ tends to zero; it follows from (III.13) and (III.14) that

$$\min \sum_{j=1}^{N} \alpha_j f_0(z_j) \geqslant \min \mu(f_0) = \eta(M_1, M_2) + \rho(\varepsilon),$$

where the minimum in the left-hand side of this inequality is over the set $P(M_1, M_2)^\varepsilon$, and the other over $Q(M_1, M_2)^\varepsilon$. The theorem follows from this expression and (III.12). □

The parameter $\varepsilon$ appearing in this theorem can be considered as the error present in numerical computations of the expressions involved in the definition of the set $P(M_1, M_2)^\varepsilon$. The theorem asserts, therefore, that the solution of the finite-dimensional linear programming problem can be used to approach a solution of the original infinite-dimensional linear program if the computation is sufficiently accurate, and the numbers $N$, $M_1$ and $M_2$, sufficiently large. The measure constructed in this manner is such that the value $\mu(f_0)$ is close to $\inf_Q \mu(f_0)$, and many of the constraints equations are satisfied closely. It is necessary to find a way of approximating the action of this measure by a trajectory–control pair; in this way, a sequence $\{p^j\}$ of such pairs will be constructed, with all the properties described in Chapter 1. This final approximation scheme will be developed in the following chapter.

**References**

Finite-dimensional linear programming: GASS [1].
Infinite-dimensional linear programming: DUFFIN [1], VERSHIK [1].
Some relevant ideas of measure theory: CHOQUET [1].

# 4
# Approximations

## 1 A first step

In this section we shall construct approximating pairs, necessary for the construction of the final approximating sequence $\{p^j\}$ with the good properties anticipated in Chapter 1. Let $\mu^* \in \mathcal{M}^+(\Omega)$ be a measure of the form

$$\mu^* = \sum_{j=1}^{N} \alpha_j \delta(z_j), \tag{IV.1}$$

where the coefficients $\alpha_j \geqslant 0, j = 1, 2, \ldots, N$, minimize the functional (III.8) over the set $P(M_1, M_2)^\varepsilon$ defined by the set of inequalities (III.9), and $z_j \in \omega$, a set dense in $\Omega$.The main difficulty in constructing a trajectory–control pair which approximates the action of this measure on the appropriate functions is that the set $\{\phi^g{}_i, i = 1, 2, \ldots, M_1\}$ can include only a finite number of functions which depend on the time only – indeed, since we have not specified otherwise, it may include only the function 1, that is, the function equal to unity for all $(t, x, u) \in \Omega$ – so that most probably the measure $\mu^*$ will not satisfy many of the last set of equalities in (I.14), that is, $\mu(f) = a_f$, $f \in C_1(\Omega)$. However, a trajectory–control pair does satisfy *all* these equalities. It could be expected that this difficulty can be resolved if the set of functions $\{\phi^g{}_i, i = 1, 2, \ldots, M_1\}$ is constructed so as to contain enough functions from a set of functions which are dependent on the time only and whose linear combinations are dense in $C_1(\Omega)$. Indeed, this turns out to be the case. It is convenient to separate these functions from the rest and relabel all the functions. The functions in the sets

$$\{f_0\}, \{\phi^g{}_i, i = 1, 2, \ldots, M_1\}, \{\chi_h, h = 1, 2, \ldots, M_2\}$$

which do not depend only on the time will be denoted as $f_k$, $k = 0, 1, \ldots, Q$; in this way we manage to retain the notation for $f_0$ – we are assuming that it does not depend on the time only. The functions in $C_1(\Omega)$ whose linear combinations are dense in this subspace will be chosen to be

$$\theta_r(t, x, u) = t^r, r = 0, 1, \ldots, \tag{IV.2}$$

and we shall assume that there are a number $L$ of them in the set $\{\phi^g_i, i = 1, 2, \ldots, M_1\}$, so that $M_1 + M_2 = Q + L$. Other functions in $C_1(\Omega)$ not of the type (IV.2) are assumed to occur later in the sequence of the $\phi_i$'s; actually, no other such function needs to be taken into account.

The measure $\mu^* \in \mathcal{M}^+(\Omega)$ satisfies

$$\begin{aligned} |\mu^*(f_k) - b_k| &\leqslant \varepsilon, k = 1, 2, \ldots, Q, \\ |\mu^*(\theta_r) - a_r| &\leqslant \varepsilon, r = 0, 1, \ldots, L, \end{aligned} \qquad \text{(IV.3)}$$

where $a_r$ is the integral of $\theta_r$ over the interval $J$, and $b_k$ stands for the corresponding value $\Delta\phi$, $k = 1, 2, \ldots, Q$.

We proceed now to construct a trajectory–control pair which approximates the action of $\mu^*$ on $f_k$, $k = 0, 1, 2, \ldots, Q$, provided that $L$ is sufficiently large. We shall assume that the integer $Q$ is fixed, while allowing $L$ to vary. This construction is a relative of one appearing in GHOUILA-HOURI [1]; the general approach was suggested by the construction appearing in this reference – in a very different framework – but the end product is not very similar; our problem is harder.

Given a number $\varepsilon_1 > 0$, it is possible to find numbers

$$t_0 = t_a < t_1 < \ldots < t_i < \ldots < t_R = t_b \qquad \text{(IV.4)}$$

and Borel sets $V_1, V_2, \ldots, V_j, \ldots, V_S$, forming a partition of $A \times U$, such that for any $i = 1, 2, \ldots, R$, $j = 1, 2, \ldots, S$, $k = 0, 1, 2, \ldots, Q$,

$$\left.\begin{array}{l} (t, t') \in [t_{i-1}, t_i) \\ (x, u), (x', u') \in V_j \end{array}\right\} \Rightarrow |f_k(t, x, u) - f_k(t', x', u')| < \varepsilon_1. \qquad \text{(IV.5)}$$

We shall assume without loss of generality that the support of the measure $\mu^*$ does not include triples $(t, x, u)$ with $t = t_i$, $i = 2, 3, \ldots, R - 1$; if it does contain such triples, the corresponding values of $t_i$ should be perturbed in such a manner that the relationships (IV.5) are maintained. It is important to note that, since the functions $f_k$, $k = 0, 1, 2, \ldots, Q$, are fixed, these partitions of $J$ and $A \times U$ are also fixed; they do not depend on $L$, which shall be allowed to vary; this variation does not then perturb the partitions in any way. Let

$$K_{ij} \equiv \mu^*([t_{i-1}, t_i) \times V_j), \qquad \text{(IV.6)}$$

and define

$$\begin{aligned} F_i(t, x, u) &= 1, \text{ if } (t, x, u) \in [t_{i-1}, t_i) \times A \times U \\ F_i(t, x, u) &= 0 \text{ otherwise.} \end{aligned}$$

Then,

$$\sum_{j=1}^{S} K_{ij} = \mu^*([t_{i-1}, t_i) \times A \times U) = \mu^*(F_i). \qquad \text{(IV.7)}$$

We note that the functions $F_i$, $i = 1, 2, \ldots, R$, are dependent on the time only, so we shall write their values as $F_i(t)$. Let $P_i^L$ be the function of time consisting of the first $L$ terms of the Chebyshev approximation of $F_i$, for $i = 2, 3, \ldots, R - 1$; for $i = 1$ it consists of the first $L$ terms of the Chebyshev approximation, not of $F_1$, but of the function $G_1$ obtained by extending $F_1$ to the part of the $t$-axis for which $t < t_a$ in such a manner that $G_1(t_a - h) = G_1(t_a + h)$ for all positive $h$ not larger than $\Delta t$. For $i = R$ we extend $F_R$ in a similar way beyond $t = t_b$, and take $P_R^L$ as the first $L$ terms in the corresponding Chebyshev approximation. We note that, for $i = 2, \ldots, R - 1$, the error functions exhibit a pulse-like behaviour near the endpoints of the corresponding time intervals – where the discontinuities occur. The amplitude of these 'pulses' does not tend to zero as $L \to \infty$, while the area of each pulse does tend to zero as $L \to \infty$, so the process can be described as a shrinkage around the point of discontinuity. Similar phenomena occur for $i = 1$, at $t_1$, and for $i = R$, at $t_{R-1}$. We have

$$P_i^L = \sum_{r=1}^{L} \beta_{ir}\theta_r.$$

Also, since $\mu^*$ satisfies the relationships (IV.3),

$$\mu^*(\theta_r) = a_r + \lambda_r(\varepsilon),$$

with $\varepsilon \to \lambda_r(\varepsilon)$ a function which tends to zero as $\varepsilon$ tends to zero. Thus,

$$\begin{aligned}\mu^*(P_i^L) &= \mu^*\left(\sum_{r=1}^{L} \beta_{ir}\theta_r\right) = \sum_{r=1}^{L} \beta_{ir}a_r + \sum_{r=1}^{L} \beta_{ir}\lambda_r(\varepsilon)\\ &= \int_J P_i^L(t)\mathrm{d}t + \sigma_i(\varepsilon)\\ &= \int_J [P_i^L(t) - F_i(t)]\mathrm{d}t + \int_J F_i(t)\mathrm{d}t + \sigma_i(\varepsilon)\\ &= (t_i - t_{i-1}) + \int_J [P_i^L(t) - F_i(t)]\mathrm{d}t + \sigma_i(\varepsilon);\end{aligned}$$

here

$$\sigma_i(\varepsilon) = \sum_{r=1}^{L} \beta_{ir}\lambda_r(\varepsilon),$$

which tends to zero as $\varepsilon$ tends to zero in a manner that depends on the value of the index $L$. We can estimate

$$\begin{aligned}\mu^*(F_i) &= \mu^*(F_i - P_i^L) + \mu^*(P_i^L)\\ &= (t_i - t_{i-1}) + \mu^*(F_i - P_i^L) + \int_J [P_i^L(t) - F_i(t)]\mathrm{d}t + \sigma_i(\varepsilon)\\ &= \Delta_i + \delta_i^L,\end{aligned}$$

where

$$\Delta_i = t_i - t_{i-1},\ \delta_i^L = \mu^*(F_i - P_i^L) + \int_J [P_i^L(t) - F_i(t)]\mathrm{d}t + \sigma_i(\varepsilon).$$

Define now

$$H_{ij} = K_{ij}(1 + \rho_i^L),$$

with $\rho_i^L$ to be defined below. Then,

$$\sum_{j=1}^{S} H_{ij} = \sum_{j=1}^{S} K_{ij} + \rho_i^L \sum_{j=1}^{S} K_{ij} = \Delta_i + \delta_i^L + \rho_i^L\Delta_i + \rho_i^L\delta_i^L = \Delta_i, \quad \text{(IV.8)}$$

since we shall choose $\rho_i^L$ so that

$$\delta_i^L + \rho_i^L\Delta_i + \rho_i^L\delta_i^L = 0,$$

that is,

$$\rho_i^L = -\delta_i^L/(\Delta_i + \delta_i^L).$$

Also,

$$K_{ij} = H_{ij}/(1 + \rho_i^L) = \zeta_i^L H_{ij}, \quad \text{(IV.9)}$$

with

$$\zeta_i^L = 1/(1 + \rho_i^L) = 1 + \delta_i^L/\Delta_i. \quad \text{(IV.10)}$$

We can proceed now to the construction of a pair $q = [x(.), u(.)]$ which approximates the action of $\mu^*$ on the functions $f_k$, $k = 0, 1, 2, \ldots, Q$. Let $(x_j, u_j)$ be an element of $V_j$, for $j = 1, 2, \ldots, S$. Define the pair $q$ as follows:

$$x(t) = x_j,\ u(t) = u_j,\ t \in B_{ij},$$

$$B_{ij} = \left(t_{i-1} + \sum_{k<j} H_{ik},\ t_{i-1} + \sum_{k\leqslant j} H_{ik}\right).$$

Since those intervals $B_{ij}$ for which $H_{ij} = 0$ are reduced to a point, they do not contribute anything to integrals such as those in the definition of the numbers $l_{ijk}$ below, and can be ignored. Instead of relabelling those intervals for which $H_{ij} > 0$, etc., we shall assume in the analysis to follow without loss of generality that $H_{ij} > 0$ for $i = 1, 2, \ldots, R$, $j = 1, 2, \ldots, S$.

If $\mu_q$ is the measure associated with the pair $q = [x(.), u(.)]$, the numbers

$$l_{ijk} = \int_{B_{ij}\times A\times U} f_k(t, x, u)\mathrm{d}\mu_q = \int_{B_{ij}} f_k(t, x_j, u_j)\mathrm{d}t$$

satisfy, for all $i, j, k$,

$$H_{ij}I_{ijk} \leqslant l_{ijk} \leqslant H_{ij}S_{ijk} \quad \text{(IV.11)}$$

where

$$I_{ijk} = \inf \{ f_k(t, x, u): (t, x, u) \in [t_{i-1}, t_i) \times V_j \}$$

$$S_{ijk} = \sup \{ f_k(t, x, u): (t, x, u) \in [t_{i-1}, t_i) \times V_j \}.$$

Also, if

$$T_{ijk} = \int_{[t_{i-1}, t_i) \times V_j} f_k \mathrm{d}\mu^*,$$

then

$$K_{ij} I_{ijk} \leqslant T_{ijk} \leqslant K_{ij} S_{ijk}. \qquad \text{(IV.12)}$$

Thus, from (IV.9), (IV.11) and (IV.12),

$$H_{ij}(I_{ijk} - \zeta_i^L S_{ijk}) \leqslant l_{ijk} - T_{ijk} \leqslant H_{ij}(S_{ijk} - \zeta_i^L I_{ijk}). \qquad \text{(IV.13)}$$

From the definition of $\zeta_i^L$ and $\delta_i^L$ it follows that if $L$ is large enough and $\varepsilon$ small enough, $\zeta_i^L$ is as near as desired to 1; in particular, under these conditions,

$$H_{ij} > 0.$$

We assume that this is the case; then,

$$|l_{ijk} - T_{ijk}| \leqslant H_{ij} \max \{ |I_{ijk} - \zeta_i^L S_{ijk}|, |S_{ijk} - \zeta_i^L I_{ijk}| \}$$
$$\leqslant H_{ij}(\varepsilon_1 + |1 - \zeta_i^L| \max (|I_{ijk}|, |S_{ijk}|)), \qquad \text{(IV.14)}$$

since

$$|I_{ijk} - \zeta_i^L S_{ijk}| \leqslant |I_{ijk} - S_{ijk}| + |1 - \zeta_i^L| \, |S_{ijk}|,$$

$$|S_{ijk} - \zeta_i^L I_{ijk}| \leqslant |S_{ijk} - I_{ijk}| + |1 - \zeta_i^L| \, |I_{ijk}|.$$

Adding all the inequalities (IV.12) with respect to $i$ and $j$, we obtain

$$\left| \int_\Omega f_k \mathrm{d}\mu^* - \int_J f_k[t, x(t), u(t)] \mathrm{d}t \right|$$
$$\leqslant \varepsilon_1 \sum_{i=1}^{R} \sum_{j=1}^{S} H_{ij} + \sum_{i=1}^{R} \sum_{j=1}^{S} H_{ij} |1 - \zeta_i^L| \max (|I_{ijk}|, |S_{ijk}|)$$
$$\leqslant \varepsilon_1 \Delta t + \varepsilon_1 \Delta t = 2 \, \varepsilon_1 \Delta t, \; k = 0, 1, 2, \ldots, Q; \qquad \text{(IV.15)}$$

here we have used (IV.8) and chosen $\varepsilon$ and $L$ so that

$$|1 - \zeta_i^L| \max (|I_{ijk}|, |S_{ijk}|) = s|\delta_i^L| \leqslant \varepsilon_1, \qquad \text{(IV.16)}$$

for $i = 1, 2, \ldots, S$; here $s$ stands for the maximum appearing in the first of these equalities, divided by $\Delta_i$. We remind the reader that

$$\delta_i^L = \mu^*(F_i - P_i^L) + \int_J [P_i^L(t) - F_i(t)] \mathrm{d}t + \sigma_i(\varepsilon),$$

so that to achieve the inequalities (IV.16) we must first choose $L$ so that each of the first two terms in this expression is not higher than $\varepsilon_1/4s$ in absolute value, for $i = 1, 2, \ldots, S$ – we are assuming without loss of generality that $s \neq 0$. We should remember the pulse-like behaviour of the error function $F_i - P_i^L$, discussed when defining $P_i^L$. Since the area of the pulses does tend to zero as $L \to \infty$, there is no doubt about the possibility of making the integral term above not higher than $\varepsilon_1/4s$ by choosing $L$ sufficiently large. With respect to the term $\mu^*(F_i - P_i^L)$, the support of $\mu^*$ does not include triples $z$ with values $t_i$, $i = 2, \ldots, R - 1$, and since the width of the pulses – which occur precisely at these points, and at no other – tends to zero as $L \to \infty$, can also be made no higher than $\varepsilon_1/4s$ by increasing $L$. Then once $L$ is chosen, we can choose $\varepsilon$ so that $|\sigma_i(\varepsilon)| \leqslant \varepsilon_1/2s$, since $|\sigma_i(\varepsilon)| \leqslant c\varepsilon$ for a constant $c$ which depends on $L$; we take

$$\varepsilon \leqslant \varepsilon_1/2sc. \tag{IV.17}$$

The parameter $\varepsilon_1$ measures the accuracy expected in this approximation. Note that in order to achieve a given accuracy the parameter $\varepsilon$ associated with the linear programming problem treated in Proposition III.3 and Theorem III.1 must be sufficiently small, and that this requirement may – probably will – influence the size of the parameter $N$ appearing in the linear program. However, the integer $Q$ is not so influenced, so all is well: if the choice of $\varepsilon$ had influenced $Q$ we would have had a rather unpleasant situation, since the required value of $\varepsilon$ does depend on $Q$, as is clear in (IV.17).

We can put together these results plus those from previous approximation schemes, and prove

**Theorem IV.1** Given $\varepsilon_2 > 0$, there exists a pair

$$q(\varepsilon_2, M_1, M_2) = [x_c(.), u(.)],$$

where $x_c(.)$ is *continuous* and $u(.)$ piecewise constant, such that

$$\begin{aligned} \eta(M_1, M_2) + \rho_1(\varepsilon_2) \leqslant \mu(\varepsilon_2, M_1, M_2)(f_0) &\leqslant \eta(M_1, M_2) + \varepsilon_2, \\ |\mu(\varepsilon_2, M_1, M_2)(\phi^g{}_i) - \Delta\phi_i| &\leqslant \varepsilon_2,\ i = 1, 2, \ldots, M_1, \\ |\mu(\varepsilon_2, M_1, M_2)(\chi_h)| &\leqslant \varepsilon_2,\ h = 1, 2, \ldots, M_2. \end{aligned} \tag{IV.18}$$

Here $\rho_1(\varepsilon_2) \to 0$ as $\varepsilon_2 \to 0$, and $\mu(\varepsilon_2, M_1, M_2)$ is the measure associated with the pair $q(\varepsilon_2, M_1, M_2)$.

*Proof* (i) Let $\delta_1 > 0$, and $v(\varepsilon_1, M_1, M_2)$ be the measure associated with the pair $q$ constructed in the analysis prior to this theorem; we called this measure temporarily $\mu_q$ in this analysis. Putting $\varepsilon_1 = \delta_1/(1 + 1/2sc)$, we find that, by Theorem III.1, (IV.3), (IV.16) and (IV.17),

$$|\nu(\varepsilon_1, M_1, M_2)(f_k) - b_k| \leqslant |\nu(\varepsilon_1, M_1, M_2)(f_k) - \mu^*(f_k)| + |\mu^*(f_k) - b_k|$$
$$\leqslant \varepsilon_1 + \varepsilon \leqslant \varepsilon_1(1 + 1/2sc) = \delta_1, \quad \text{(IV.19)}$$

for $k = 1, 2, \ldots, Q$; a similar analysis can be made for the inequalities corresponding to the function $f_0$. The measure $\nu(\varepsilon_1, M_1, M_2)$ satisfies exactly, of course, those equalities of the type $\mu(f) = a_f, f \in C_1(\Omega)$.

(ii) We have to show that the trajectory function $x(.)$ in the pair $q$ constructed above can be modified so as to construct a continuous function $x_c(.)$, rather than a piecewise constant one, so that the pair $q(\varepsilon_2, M_1, M_2) = [x_c(.), u(.)]$ satisfies the relationships (IV.18), for a given $\varepsilon_2 > 0$. The piecewise constant function $x(.)$ has (possibly) points of discontinuity at the endpoints of the interval $B_{ij}$; call these points $t_{ij}$ and $t_{i(j+1)}$, so that $B_{ij} = [t_{ij}, t_{i(j+1)})$; then $t_{i-1} = t_{i1}$ and $t_i = t_{i(S+1)}$. On $B_{i(j-1)}$ we have $x(t) = x_{j-1}$ and on $B_{ij}$ we have $x(t) = x_j$; these two values may be different, so that $x(.)$ has a discontinuity at $t_{ij}$. We proceed to construct $x_c(.)$ from $x(.)$ as follows. Let

$$I_{ij} = [t_{ij} - a_{ij}^-, t_{ij} + a_{ij}^+],$$

where the numbers $a_{ij}^-$ and $a_{ij}^+$ are chosen so that

$$[t_{ij} - a_{ij}^-, t_{ij}] \subset (\tau_{i(j-1)}, t_{ij}]$$
$$[t_{ij}, t_{ij} + a_{ij}^+] \subset [t_{ij}, \tau_{ij}),$$

where $\tau_{i(j-1)}$, $\tau_{ij}$ are the midpoints of $B_{i(j-1)}$ and $B_{ij}$ respectively. As a matter of fact, we shall assume without loss of generality that $\delta_1 < 1/2$, and choose

$$|a_{ij}^+| = \delta_1 |B_{ij}|, |a_{ij}^-| = \delta_1 |B_{i(j-1)}|,$$

where $|.|$ indicates the length of an interval. Define $x_c(.)$ as $x_{j-1}$ on $[t_{i(j-1)}, t_{ij} - a_{ij}^-)$ and as $x_j$ on $[t_{ij} + a_{ij}^+, t_{i(j+1)})$. Since the set $A$ is pathwise connected (see Section 2 of Chapter 1), it is possible to find a path $\theta \to X_{ij}(\theta)$, $\theta \in [0, 1]$, such that $X_{ij}(0) = x_{j-1}$, $X_{ij}(1) = x_j$, and $X_{ij}(.)$ is continuous. We map $I_{ij}$ into $[0, 1]$ by a continuous monotonic function

$$t \to \xi_{ij}(t) \in [0, 1], t \in I_{ij},$$

so that

$$\xi(t_{ij} - a_{ij}^-) = 0, \xi(t_{ij} + a_{ij}^+) = 1,$$

and put

$$x_c(t) = X_{ij}[\xi_{ij}(t)] \text{ on } I_{ij}.$$

Then $x_c(.)$ is continuous on $J$. Let $\mu$ be a temporary name for the measure associated with the pair $[x_c(.), u(.)]$,

$$
\begin{aligned}
|\nu(\varepsilon_1, M_1, M_2)(f_k) - \mu(f_k)| &= \left| \int_J \{f_k[t, x(t), u(t)] - f_k[t, x_c(t), u(t)]\}\mathrm{d}t \right| \\
&= \left| \sum_{i,j} \int_{B_{ij}} \{f_k[t, x_j, u_j] - f_k[t, x_c(t), u_j]\}\mathrm{d}t \right| \\
&= \left| \sum_{i,j} \int_{I_{ij}} \{f_k[t, x_j, u_j] - f_k[t, x_c(t), u_j]\}\mathrm{d}t \right| \\
&\leqslant 2w \sum_{i,j} |I_{ij}| \leqslant 4\delta_1 w \sum_{i,j} |B_{ij}| \\
&= 4w\delta_1 \Delta t, \; k = 0, 1, 2, \ldots, Q, \qquad \text{(IV.20)}
\end{aligned}
$$

where

$$w = \max_{0 \leqslant k \leqslant Q} \sup_{\Omega} |f_k(t, x, u)|.$$

Finally, from (IV.19) and (IV.20),

$$|\mu(f_k) - b_k| \leqslant \delta_1(1 + 4w\Delta t), \; k = 0, 1, 2, \ldots, Q,$$

so that if we choose $\delta_1 \leqslant \varepsilon_2/(1 + 4w\Delta t)$, the measure $\mu$, to be relabelled $\mu(\varepsilon_2, M_1, M_2)$, satisfies the inequalities (IV.18). □

We have been successful in constructing pairs which approximate well – as well as desired – the action on the appropriate functions of the optimal measure constructed by solving the linear programming problem of Theorem III.1.

It should be noted that these pairs do depend on the numbers $M_1$ and $M_2$ of functions and inequalities appearing in the linear program; we have not made this fact explicit in the notation so that it does not become unwieldy. Of course, the results of Proposition III.1 tell us that if we increase these numbers $M_1$ and $M_2$ the infimum $\eta(M_1, M_2)$ approaches the infimum $\eta$; in other words, if the number $\varepsilon_2$ is small and the number of functions, or of equations, $M_1$ and $M_2$, are large enough, we have a trajectory–control pair which assigns to the performance criterion a value near $\eta$, and which nearly satisfies – to a high degree of accuracy – many of the equalities of the type $\mu(\phi) = \Delta\phi$. We note that if the number $\varepsilon_2$ is small, the parameter $N$ in the linear program probably needs to be large, if accuracy is to be maintained; if $\varepsilon_2$ is small, according to the analysis before and during the proof of Theorem IV.1 the parameter $\varepsilon$ should be small, which probably implies that the number of unknowns $N$ in the linear program (III.8) and (III.9) should be large; see especially the first few paragraphs of the proof of Theorem III.1.

It is not quite clear whether such trajectory–control pairs, in spite of their interesting characteristics, can be considered as a solution to our modified control problem, described in the last section of Chapter 1. The problem seems to arise out of the nature of the trajectory function. Is it really a trajectory, that is, the response of the controlled system to an admissible control? This is really the *key question*. Of course, other questions should be asked. Does this function satisfy the boundary conditions? Are the points $x_c(t)$, $t \in J$, contained in the set $A$? The control function $u(.)$ is, however, admissible – it obviously takes values in the set $U$, and is piecewice continuous and bounded, thus summable. As a matter of fact, the final construction of approximating pairs will consist of pairs in which the trajectory function will be the response, call it $x_R(.)$, of the controlled system to this control function $u(.)$, with initial condition $x_R(t_a) = x_a$. These pairs, of the form $[x_R(.), u(.)]$, turn out to be the solution to the modified control problem; we note that the functions $g$ and $f_0$ will be required to be Lipschitz, rather than merely continuous, for these properties to hold.

## 2 The solution to the modified control problem

We shall prove further on in this section that, for appropriate values of the appropriate indices, provided that the functions $g$ and $f_0$ are Lipschitz, the values of the functions $x_c(.)$ and $x_R(.)$ are close; in this way we shall be able to answer questions concerning the trajectory functions $x_R(.)$, simply by deriving properties of the functions $x_c(.)$. For instance, it follows from the construction of these functions that they take their values in the constraint set $A$ – thus the values of the functions $x_R(.)$ cannot stray very far from this set, and indeed the distances from the graphs of these functions to the set $A$ can be made as close to zero as may be required.

It is necessary, then, to determine certain properties of the pairs $q(\varepsilon_2, M_1, M_2)$ of Theorem IV.1. We assume in all that follows that the functions $f_0$ and $g$ are Lipschitz in $\Omega$, that is, that constants $k_0$ and $k_g$ exist such that

$$|f_0(t', x', u') - f_0(t, x, u)| \leqslant k_0(|t' - t| + \|x' - x\| + \|u' - u\|),$$
$$\|g(t', x', u') - g(t, x, u)\| \leqslant k_g(|t' - t| + \|x' - x\| + \|u' - u\|), \tag{IV.21}$$

for all $(t', x', u'), (t, x, u) \in \Omega$; the same symbol is used for the euclidean norms in $R^n$, $R^m$ and, below, in $R^{n+1}$.

**Proposition IV.1** The pairs $q(\varepsilon_2, M_1, M_2) = [x_c(.), u(.)]$ of Theorem IV.1, have the following properties:

(i) $x_c(t) \in A$, $t \in J$.
(ii) $u(t) \in U$, $t \in J$.
(iii) Given $\delta > 0$, the numbers $\varepsilon_2$ and $M_1$ can be chosen such that

$$\|x_c(t_a) - x_a\| \leqslant \delta, \; \|x_c(t_b) - x_b\| \leqslant \delta;$$

it may be necessary to modify the trajectory function $x_c(.)$ near $t_a$ and $t_b$ to achieve this property.

*Proof* The properties (i) and (ii) follow readily from the construction of these functions. The proof of the property (iii) can be best presented by dividing it into several stages.

(a) It is necessary to choose once and for all the functions $\phi_i$ in (IV.18); they are to be $M_1$ monomials in $t$ and the components of $x$. We separate them first into classes, the first class being that of first-degree monomials, the second that of second-degree ones, etc. In each class we follow a cyclic ordering based on the indices of the components of the $(n + 1)$-tuple $(t, x)$; note that the time $t$ is the first component, $x_1$ the second, $x_2$ the third, . . ., $x_n$ the $(n + 1)$st. We note that this ordering does impose an extra constraint on the index $M_1$, in the sense that, if the index $L$ of the analysis prior to Theorem IV.1 is chosen, perhaps because the number $\varepsilon_1$ has been chosen, or determined by other considerations, then the index $M_1$ must be not smaller than a value $M_1'(L)$.

We remind the reader that the set $J \times A$ is contained in an open ball $B$ in

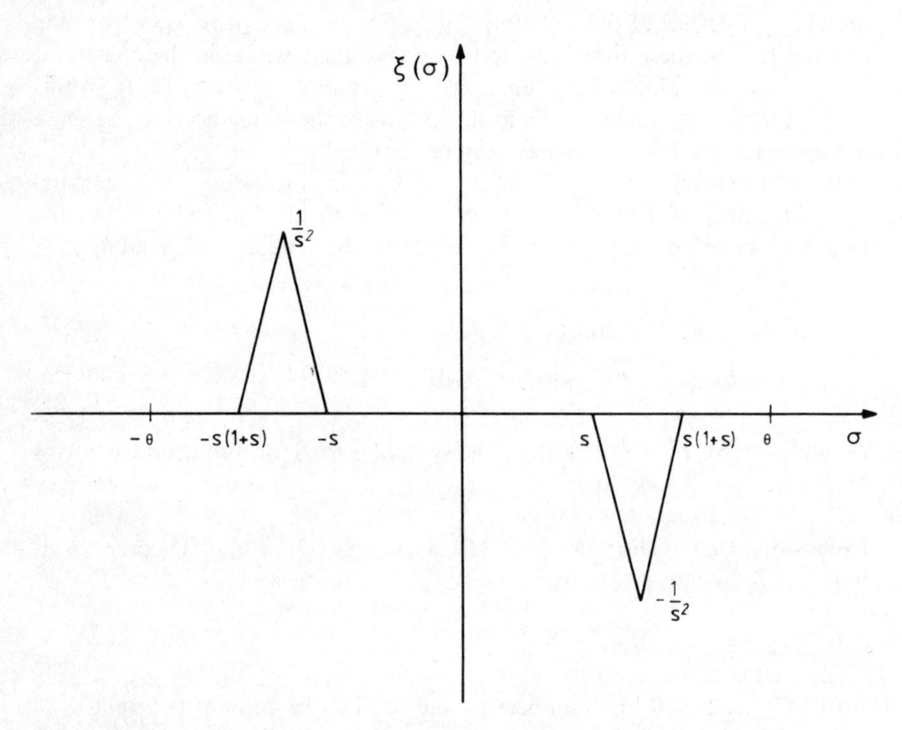


Fig. 1. Graph of the function $\sigma \to \xi(\sigma)$, $\sigma \in [\theta, -\theta]$, used in the construction of some polynomials.

$R^{n+1}$; let us define this further, and say that this ball has centre at the origin of this space, and radius $\theta$. Without loss of generality – by increasing the radius $\theta$ if necessary – we shall assume that all points of $J \times A$ are at least at a distance $\theta/2$ from the boundary of $B$.

(b) Consider the odd function $\sigma \to \xi(\sigma)$, $\sigma \in [\theta, -\theta]$, whose graph is shown in Figure 1; $s$ is a parameter to be chosen below. Consider the problem of approximating this function over the interval $[-\theta, \theta]$ by an odd polynomial of degree $D$, say, in $\sigma$. By Jackson's theorem (see, for instance, LORENTZ [1]), a polynomial $p$ exists such that

$$\max_{|\sigma|<\theta} |p(\sigma) - \xi(\sigma)| \leqslant K_1/s^4 D, \tag{IV.22}$$

with $K_1$ dependent only on $\theta$. Put for $\sigma \in [-\theta, \theta]$

$$p_1(\sigma) = \int_{-\theta}^{\sigma} p(y)\mathrm{d}y, \; \xi_1(\sigma) = \int_{-\theta}^{\sigma} \xi(y)\mathrm{d}y; \tag{IV.23}$$

then there is a constant $K_2$, depending only on $\theta$, such that

$$\max_{|\sigma|\leqslant\theta} |p_1(\sigma) - \xi_1(\sigma)| \leqslant K_2/s^4 D; \tag{IV.24}$$

the even polynomial $p_1$ is of degree $D + 1$, which is determined by the index $M_1$. Let $R = R(M_1)$ be such that the monomials $t^{2R}$, $x_1^{2R}$, $\ldots$, $x_n^{2R}$, are included in the first $M_1$ monomials, but such that not all monomials $t^{2R+2}$, $x_1^{2R+2}$, $\ldots$, $x_n^{2R+2}$, are included; then $D(M_1) + 1 = 2R(M_1)$, which defines a nondecreasing step function $M_1 \to D(M_1)$. Of course, $D(M_1) \to \infty$ as $M_1 \to \infty$. We sometimes write $D$ for $D(M_1)$.

(c) In the definition of the function $\xi$, put $s = D^{-1/8}$. Then $s(1 + s) \to 0$ as $M_2 \to \infty$, and a neighbourhood in $R^{n+1}$ of radius $s(1 + s)$ of any point in $J \times A$ can be made to be wholly inside the ball $B$ by choosing $M_1$ sufficiently high. Also, let

$$G_a = \{(t, x)\colon \|(t, x) - (t_a, x_a)\| \leqslant s(1 + s), (t, x) \in J \times A\},$$

and choose $M_1 > M'$, say, so that

$$s(s + 1) \leqslant \min [\theta/2, \|(t_b, x_b) - (t_a, x_a)\|];$$

then $(t_b, x_b)$ is not in $G_a$ for $M_1 > M'$; here the norm is euclidean in $R^{n+1}$. Assume that the trajectory $x_c(.)$, constructed as in Theorem IV.1, is wholly outside this set. We shall prove that, for the index $M_1$ sufficiently large and the number $\varepsilon_2$ sufficiently small, this assumption will lead us to a contradiction; the trajectory $x_c(.)$ and the set $G_a$ – a neighbourhood of $(t_a, x_a)$ in $J \times A$ – intersect.

Indeed, let $\mu_q$ be the measure associated with the pair $q(\varepsilon_2, M_1, M_2)$. Then, according to Theorem IV.1, this measure, or pair, satisfies

$$|\mu_q(\phi^g{}_i) - \Delta\phi_i| \leqslant \varepsilon_2 \tag{IV.25}$$

for the first $M_1$ monomials in our ordering. This implies that the action of $\mu_q$ on polynomials can be estimated. Put

$$\phi(t, x) = p_1[\|(t, x) - (t_a, x_a)\|]$$
$$= \sum_{i=1}^{M_1} \beta_i \phi_i(t, x), \; (t, x) \in B, \tag{IV.26}$$

for some coefficients $\beta_i$, $i = 1, 2, \ldots, M_1$. Then,

$$\|\text{grad } \phi(t, x)\| = p[\|(t, x) - (t_a, x_a)\|], \; (t, x) \in B,$$

as can be ascertained easily; the vector grad $\phi$ has as components the derivatives $\phi_t, \phi_{x1}, \ldots, \phi_{xn}$. We can make use now of our assumption – that $x_c(.)$ is wholly outside the set $G_a$ – to estimate

$$|\mu_q(\phi^g) - \Delta\phi| = \left| \int_J \phi^g[t, x_c(t), u(t)]\mathrm{d}t - \Delta\phi \right|.$$

By (IV.22), since $\xi(\sigma) = 0$ for $|\sigma| > s(1 + s)$,

$$\|\text{grad } \phi[t, x_c(t)]\| \leqslant K_1/s^4 D = K_1 D^{-1/2}, \tag{IV.27}$$

since we have chosen $s = D^{-1/8}$. Since $\phi^g = g\phi_x + \phi_t$,

$$|\mu_q(\phi^g)| \leqslant \Delta t K_1 K_3 D^{-1/2}, \tag{IV.28}$$

where

$$K_3 = \max_{\Omega} \|[1, g(t, x, u)]\|;$$

the vector $[1, g(t, x, u)]$ has components $1, g_1(t, x, u), \ldots, g_n(t, x, u)$.

Also, since $(t_b, x_b)$ is outside the set $G_a$ – we are already assuming that $M_1 > M'$ – (IV.24) implies that

$$-1 - 2K_2 D^{-1/2} \leqslant \Delta\phi = \phi_b - \phi_a \leqslant -1 + 2K_2 D^{-1/2}. \tag{IV.29}$$

Thus, for $M_1 > M'$, (IV. 28) and (IV. 29) imply that

$$|\mu_q - \Delta\phi - 1| \leqslant (\Delta t K_1 K_3 + 2K_2) D^{-1/2}; \tag{IV.30}$$

that is, $\mu(\phi^g) - \Delta\phi$ is very near 1 for sufficiently large values of $M_1$; remember that $D = D(M_1) \to \infty$ as $M_1 \to \infty$. We show that, for sufficiently small values of the number $\varepsilon_2$ this contradicts another estimation that can be made of this same quantity. Indeed, (IV.25) and (IV.26) imply that

$$|\mu_q(\phi^g) - \Delta\phi| \leqslant \varepsilon_2 \sum_{i=1}^{M_1} |\beta_i|; \tag{IV.31}$$

if we choose $M_1 \geqslant M'$ and $\varepsilon_2 = \varepsilon_2(M_1)$ such that

$$\varepsilon_2 < [1 - (\Delta t K_1 K_3 + 2K_2) D^{-1/2}] \Big/ \sum_{i=1}^{M_1} |\beta_i|, \tag{IV.32}$$

then the two inequalities, (IV.30) and (IV.31), are not compatible. Thus, a contradiction has occurred, and our only assumption, that the trajectory $x_c(.)$ does not intersect the set $G_a$, must be incorrect for the range of values of $M_1$ and $\varepsilon_2$ described above.

(d) Since $\varepsilon_2 \to \delta_1 \to \varepsilon_1 \to \varepsilon \to L \to M_1$, we call the value of $M_1$ thus chosen, $M_1''(L)$, and we choose $M_1$ now so that $s(1 + s) = D^{-1/8}(1 + D^{-1/8}) < \delta$ and $M_1 \geqslant \max [M', M_1'(L), M_1''(L)]$; choosing then the number $\varepsilon_2$ so as to satisfy (IV.32), we conclude that the trajectory $x_c(.)$ intersects a neighbourhood $G_a$ in $J \times A$ of the point $(t_a, x_a)$, itself contained in a ball of radius $\delta$. This is not, alas, enough to guarantee that $x(t_a)$ itself is in this neighbourhood; it may be that the trajectory $x_c(.)$ leaves $G_a$ for values of the time very close to $t_a$, and takes the value $x(t_a)$ outside this set; it could even be that, as $\delta$ is made to tend to zero, $x(t_a)$ stays fixed, away from $x_a$. If the trajectories corresponding to different values of $\delta$ were to converge, as $\delta \to 0$ – not, perhaps, a normal occurrence – the limiting function would be discontinuous at $t_a$, under these conditions.

It is possible to modify the function $x_c(.)$ near $t_a$ so that we can indeed say that $\| x_c(t_a) - x_a \| \leqslant \delta$. Let $[t^\alpha, x^\alpha]$ be a point of the neighbourhood $G_a$ such that $x_c(t^\alpha) = x^\alpha$; we modify the function $x_c(.)$ by requiring that

$$x_c(t) = x^\alpha, \ t \in [t_a, t^\alpha]; \qquad \text{(IV.33)}$$

if $x_c(t_a)$ is in $G_a$ it can of course be taken as $x^\alpha$, and no change is required. Exactly the same analysis can be done near $x_b$, at the final end of the trajectory; the choice of $M_1$ and $\varepsilon_2$ above ensures also that the unmodified trajectory $x_c(.)$ intersects a set $G_b$, defined in a way similar to $G_a$ but centred at $x_b$ rather than at $x_a$. Again, it may be necessary to modify the trajectory near $t_b$, by choosing a point $(t^\beta, x^\beta) \in G_b$ such that $x_c(t^\beta) = x^\beta$, and putting

$$x_c(t) = x^\beta, \ t \in [t^\beta, t_a]; \qquad \text{(IV.34)}$$

this will ensure that $\| x_c(t_b) - x_b \| < \delta$. We note that these modifications may have some unintended effects. First, the modification near $t_a$ will influence the behaviour near $t_b$; we leave to the reader to show that this effect is negligible with respect to our conclusions above. Also, the values that the pair $[x_c(.), u(.)]$ assigns to the functions in (IV.18), Theorem IV.1, may change with these modifications of the trajectory function $x_c(.)$; these matters need a thorough study, to be presented below. □

We start now the analysis leading to the determination of some properties of the pair $[x_R(.), u(.)]$, where, it should be remembered, the trajectory $x_R(.)$ is the response of the system (I.1) to the control function $u(.)$ constructed above, with $x(t_a) = x_a$. As announced at the beginning of this section, we shall show that the values of the two functions, $x_R(.)$ and the modified $x_c(.)$ of the last proposition, are close.

Consider the functions appearing in (IV.18), Theorem IV.1. All these functions – $f_0$, the $\phi^g{}_i$'s, the $\chi_h$'s – are, according to our assumption (IV. 21), Lipschitz in $\Omega$, since $f_0$ and $g$ are Lipschitz, and the functions $\phi$, $\psi$ used in constructing some of those functions are uniformly continuously differentiable on $\Omega$. Let then $F$ be a function which is Lipschitz on $\Omega$ with constant $k$. The actions of the original and modified trajectory functions, to be called $x_c'(.)$ and $x_c(.)$ respectively, can be compared:

$$\begin{aligned}
&\left|\int_J \{F[t, x_c'(t), u(t)] - F[t, x_c(t), u(t)]\}\mathrm{d}t\right| \\
&\leqslant \left|\int_{t_a}^{t^\alpha} \{F[t, x_c'(t), u(t)] - F[t, x^\alpha, u(t)]\}\mathrm{d}t\right| \\
&+ \left|\int_{t^\beta}^{t_b} \{F[t, x_c'(t), u(t)] - F[t, x^\beta, u(t)]\}\mathrm{d}t\right| \\
&\leqslant (t^\alpha - t_a)k \max_{[t^\alpha, t_a]} \|x_c'(t) - x^\alpha\| + (t_b - t^\beta)k \max_{[t^\beta, t_b]} \|x_c'(t) - x^\beta\| \\
&\leqslant [(t^\alpha - t_a) + (t_b - t^\beta)]k2\theta \\
&\leqslant 4\delta\theta k, \qquad \text{(IV.35)}
\end{aligned}$$

where $\theta$ is, we remind the reader, the radius of the ball $B$ containing $J \times A$.

Let us consider in detail this and related estimations for the functions $\chi_h$, $h = 1, 2, \ldots, M_2$; we remind the reader that these functions are of the form $\psi_i(t, x, u) = x_j\psi'(t) + g_j(t, x, u)\psi(t)$, when the functions $\psi$ are the sine and cosine functions (III.3), and $j = 1, 2, \ldots, n$. Suppose now that the index $M_2$ is of the form

$$M_2 = 2M_{21}n$$

so that the $M_2$ functions $\chi_h$ are the functions $\psi_j^r(t, x, u) \equiv x_j\psi_r'(t) + g_j(t, x, u)\psi_r(t)$, $r = 1, 2, \ldots, 2M_{21}$, $j = 1, 2, \ldots, n$, $(t, x, u) \in \Omega$; here

$$\begin{aligned}
&\psi_r(t) = \sin\,[2\pi r(t - t_a)/\Delta t],\ r = 1, 2, \ldots, M_{21}, \\
&\psi_r(t) = 1 - \cos\,[2\pi(r - M_{21})(t - t_a)/\Delta t],\ r = M_{21} + 1, M_{21} + 2, \ldots, 2M_{21}.
\end{aligned} \qquad \text{(IV.36)}$$

It follows from the structure of the functions $\psi_j^r$ that their Lipschitz constants, to be denoted by $k_j^r$, satisfy a relationship of the following type:

$k_j^r \leqslant c\rho$, $j = 1, 2, \ldots, n$, $r = 1, 2, \ldots, 2M_{21}$, with $\rho = r$, $r \leqslant M_{21}$, $\rho = r - M_{21}$, $r > M_{21}$.

where $c$ is a constant independent of $j$ and $r$. Thus, the estimate (IV.35) and the inequality (IV.18), Theorem IV.1, indicate that the pair $q = [x_c(.), u(.)]$, where $x_c(.)$ is, we remind the reader, the modified trajectory, assigns to the functions $\psi_j^r$ values which satisfy the following inequalities:

$$|\mu_q(\psi_j^r)| \leqslant \varepsilon_2 + 4\delta\theta c\rho,\ r = 1, 2, \ldots, 2M_{21},\ j = 1, 2, \ldots, n;$$

$\mu_q$ is the measure corresponding to the pair $q$. Thus,

$$\left| \int_J \{x_{cj}(t)\psi_r'(t) + g_j[t, x_c(t), u(t)]\psi_r(t)\}\mathrm{d}t \right|$$

$$= \left| \int_J y_j(t)\psi_r'(t)\mathrm{d}t \right|$$

$$\leqslant \varepsilon_2 + 4\delta\theta c\rho,\ r = 1, 2, \ldots, 2M_{21},\ j = 1, 2, \ldots, n; \qquad \text{(IV.37)}$$

here

$$y_j(t) = x_{cj}(t) - \int_{t_a}^{t} g_j[\tau, x_c(\tau), u(\tau)]\mathrm{d}\tau.$$

Since the pair $[x_R(.), u(.)]$ is an exact solution of the differential equation (I.1),

$$z_j(t) \equiv x_{Rj}(t) - \int_{t_a}^{t} g_j[\tau, x_R(\tau), u(\tau)]\mathrm{d}\tau = 0,\ t \in J,$$

so that $w_j(.) \equiv y_j(.) - z_j(.)$ satisfies

$$\left| \int_J w_j(t)\psi_r'(t)\mathrm{d}t \right| \leqslant \varepsilon_2 + 4\delta\theta cr$$

that is

$$\left| \int_J w_j(t) \sin\,[2\pi r(t - t_a)/\Delta t]\mathrm{d}t \right| \leqslant \varepsilon_2/r + 4\delta\theta c$$

$$\left| \int_J w_j(t) \cos\,[2\pi r(t - t_a)/\Delta t]\mathrm{d}t \right| \leqslant \varepsilon_2/r + 4\delta\theta c,$$

for $r = 1, 2, \ldots, M_{21}$. Thus, the first $2M_{21}$ Fourier coefficients of $w_j(.)$, except for the constant term, are bounded tightly, if $\varepsilon_2$ and $\delta$ are small enough; also, the tail $h_j(.)$ of the Fourier series tends to zero as $M_{21}$ – that is, $M_2$ – tends to infinity since $w_j(.)$ is continuous on $J$. We define $n$-vector-valued functions $w(.)$, $h(.)$, $y(.)$, $z(.)$, in the obvious way, and write

$$w(t) = v + \gamma(t),\ t \in J, \qquad \text{(IV.39)}$$

where $v$ is a constant $n$-vector and the euclidean norm of $\gamma(t)$, for all $t \in J$, can be made no higher than a given number $\delta_2 > 0$ by choosing $\varepsilon_2$ and $\delta$ sufficiently small, and $M_{21}$ sufficiently large. Indeed,

$$\|\gamma(t)\| \leqslant 2M_{21}n(\varepsilon_2 + 4\delta\theta c) + \|h(t)\|,$$

so that, given $\delta_2 > 0$, we first choose $M_{21}$ so that

$$\|h(t)\| \leqslant \delta_2/2,\ t \in J, \qquad \text{(IV.40)}$$

then $\varepsilon_2$ and $\delta$ so that

$$2M_{21}n(\varepsilon_2 + 4\delta\theta c) \leqslant \delta_2/2. \tag{IV.41}$$

Thus,

$$\| \gamma(t) \| \leqslant \delta_2,\ t \in J. \tag{IV.42}$$

It is necessary to relate these choices to our previous developments, since there may not be absolute freedom to make these choices because of previous commitments. It is very important to remark that in the analysis leading to the proof of Proposition IV.1 no bound was put on the index $M_2$ – only on $M_1$, so $M_{21}$ can be chosen independently of the parameter $\delta$ appearing in Proposition IV.1 as well as in the present development. A relationship *was* established between $\delta$ and $\varepsilon_2$, via the index $M_1$; no matter, this establishes only an upper bound for $\varepsilon_2$. In the proof of Theorem IV.2 below these matters will be treated in detail.

We must estimate the magnitude of the constant vector $v$ in (IV.39). Write this formula in detail as

$$\begin{aligned} w(t) &= y(t) - z(t) \\ &= x_c(t) - \int_{t_a}^{t} g[\tau, x_c(\tau), u(\tau)]\mathrm{d}\tau - x_R(t) + \int_{t_a}^{t} g[\tau, x_R(\tau), u(\tau)]\mathrm{d}\tau \\ &= v + \gamma(t),\ t \in J, \end{aligned} \tag{IV.43}$$

and put $t = t_a$ in this equality. Then,

$$x_c(t_a) = x_R(t_a) + v + \gamma(t_a);$$

that is, since $x_R(t_a) = x_a$,

$$\|v\| < \|x_c(t_a) - x_a\| + \|\gamma(t_a)\| \leqslant \delta + \delta_2, \tag{IV.44}$$

provided that $M_{21}$ and $\delta$ are chosen as in (IV.40) and (IV.41).

We have proved, then, that the values of $y(.)$ and $z(.)$ are close; this, because of the Lipschitzian character of the function $g$ and by the judicious application of Gronwall's lemma, will be shown to imply the closedness of the functions $x_c(.)$ and $x_R(.)$. Indeed, it follows from (IV.43) that

$$\|x_c(t) - x_R(t)\| = \left\| \int_{t_a}^{t} g[\tau, x_c(\tau), u(\tau)]\mathrm{d}\tau - \int_{t_a}^{t} g[\tau, x_R(\tau), u(\tau)]\mathrm{d}\tau + v + \gamma(t) \right\|$$

$$\leqslant k_g \int_{t_a}^{t} \|x_c(\tau) - x_R(\tau)\|\mathrm{d}\tau + \|v\| + \|\gamma(t)\|,$$

which implies, by a simple modification of Gronwall's lemma (see, for instance, LEFSCHETZ [1], Chapter II), that

$$\|x_c(t) - x_R(t)\| \leqslant [\|v\| + \|\gamma(t)\|] \exp(k_g \Delta t),$$

for $t \in J$, so that, if the parameters $M_{21}$ and $\delta$ are chosen as in (IV.40) and (IV.41),

$$\|x_c(t) - x_R(t)\| \leqslant (\delta + 2\delta_2) \exp (k_g \Delta t). \tag{IV.45}$$

The values of the functions $x_c(.)$ and $x_R(.)$ can indeed be made to be as close as desired on the interval $J$ by choosing the parameters in an appropriate manner.

It is not difficult now to obtain properties of the pair $[x_R(.), u(.)]$; we summarize them in the following theorem which, together with its corollary, are our final and most important result on approximation.

**Theorem IV.2** Given $\lambda > 0$, it is possible to choose the parameters $\varepsilon$, $M_1$ and $M_2$ of Theorem III.1 such that the corresponding pair $p = [x_R(.), u(.)]$ satisfies

$$\begin{gathered} \eta - \rho_2(\lambda) \leqslant \mu_p(f_0) \leqslant \eta + \lambda \\ \|x_R(t_b) - x_b\| \leqslant \lambda \\ d[x_R(t), A] \leqslant \lambda,\ t \in J, \end{gathered} \tag{IV.47}$$

where $\mu_p$ is the measure associated with the pair $p$, $\rho_2(\lambda) \to 0$ as $\lambda \to 0$; $d(x, A)$ is the distance between the point $x \in R^n$ and the set $A \subset R^n$, and, as before $\eta = \inf_Q \mu(f_0)$.

*Proof* The pair $q(\varepsilon_2, M_1, M_2)$ of Theorem IV.1 satisfies

$$\eta(M_1, M_2) + \rho_1(\varepsilon_2) \leqslant \mu(\varepsilon_2, M_1, M_2)(f_0) \leqslant \eta(M_1, M_2) + \varepsilon_2,$$

where $\rho_1(\varepsilon_2)$ tends to zero as $\varepsilon_2$ tends to zero, and $\mu(\varepsilon_2, M_1, M_2)$ is the measure associated with the pair $q(\varepsilon_2, M_1, M_2)$. As the trajectory $x_c(.)$ associated with this pair was modified in the proof of Proposition IV.1, the values assigned to the function $f_0$ may change; this change was estimated for any Lipschitzian function in (IV.35). Also, the modified function $x_c(.)$ and the function $x_R(.)$ are close, as estimated by (IV.45), so that, putting these two effects together, and noting that $f_0$ is Lipschitz with constant $k_0$,

$$|\mu_p(f_0) - \mu(\varepsilon_2, M_1, M_2)(f_0)| \leqslant 4\delta\theta k_0 + k_0 \Delta t(\delta + 2\delta_2) \exp (k_g \Delta t);$$

also, according to Proposition III.1, given $\delta_3 > 0$, the value $\eta(M_1, M_2)$ can be made to satisfy

$$|\eta - \eta(M_1, M_2)| \leqslant \delta_3$$

by taking $M_1$ and $M_2$ sufficiently large. Putting all these relationships together,

$$\begin{aligned} &\eta + \rho_1(\varepsilon_2) - \delta_3 - [4\delta\theta k_0 + k_0 \Delta t(\delta + 2\delta_2) \exp (k_g \Delta t)] \leqslant \mu_p(f_0) \\ &\leqslant \eta + \varepsilon_2 + [4\delta\theta k_0 + k_0 \Delta t(\delta + 2\delta_2) \exp k_g \Delta t)]. \end{aligned}$$

We can now choose the parameters of the problem – $M_1$, $M_2$, $\varepsilon$, $\varepsilon_1$, $\varepsilon_2$, $\delta_1$, $\delta_2$, $\delta_3$, $\delta$ – so that the contentions of the theorem are satisfied. Of course, we must take into account all the interactions between these parameters; in doing so, we shall not simply define a range of values for each parameter, but actually chose a fixed value for it in this range. The first contention is satisfied if we make

$\delta_3 \leqslant \lambda/3$, by making $M_1 = M_1{}^0$, $M_2 = M_2{}^0$;

$\varepsilon_2 \leqslant \lambda/3$, by making $\varepsilon_2 = \varepsilon_2{}^0$, which, as in Theorem IV.1, determines $\delta_1$, which determines $\varepsilon_1$, which determines $\varepsilon = \varepsilon^0$;

$4\delta\theta k_0 \leqslant \lambda/6$, by making $\delta = \delta^0$, which determines $\varepsilon_2 = \varepsilon_2{}^1$ (which, as in Theorem IV.1, determines $\delta_1$, which determines $\varepsilon_1$, which determines $\varepsilon = \varepsilon^1$) and $M_1 = M_1{}^1$, as in the proof of Proposition IV.1;

$k_0 \Delta t \delta \exp (k_g \Delta t) \leqslant \lambda/12$, by making $\delta = \delta^1$ which determines $\varepsilon_2 = \varepsilon_2{}^2$ (which, as in Theorem IV.1, determines $\delta_1$, which determines $\varepsilon_1$, which determines $\varepsilon = \varepsilon^2$) and $M_1 = M_1{}^2$, as in the proof of Proposition IV.1;

$2k_0 \Delta t \lambda_1 \exp (k_g \Delta t) \leqslant \lambda/12$, by making $\delta_2 = \delta_2{}^0$, which determines $M_2 = M_2{}^1$, $\delta = \delta^2$ (which determines $\varepsilon_2 = \varepsilon_2{}^3$ – which, as in Theorem IV.1, determines $\delta_1$, which determines $\varepsilon_1$, which determines $\varepsilon = \varepsilon^3$ – and $M_1 = M_1{}^3$, as in the proof of Proposition IV.1) and $\varepsilon_2 = \varepsilon_2{}^4$, which determines $\varepsilon = \varepsilon^4$.

With respect to the other contentions of the present theorem,

$$\begin{aligned} \|x_R(t_b) - x_b\| &\leqslant \|x_R(t_b) - x_c(t_b)\| + \|x_c(t_b) - x_b\| \\ &\leqslant (\delta + 2\delta_2) \exp (k_g \Delta t) + \delta; \end{aligned} \qquad \text{(IV.48)}$$

also, since the vectors $\{x_c(t), t \in J\}$ are in the set $A$, and the difference between the values of $x_c(.)$ and $x_R(.)$ satisfies (IV.45),

$$d[x_R(t), A] \leqslant (\delta + 2\delta_2) \exp (k_g \Delta t), \ t \in J. \qquad \text{(IV.49)}$$

The right-hand side of (IV.49) is not higher that $\lambda/6$, because of our choices above, so it need not concern us any more; the third contention of the theorem is then satisfied. The right-hand side of (IV.48) is not higher than $(\lambda/6) + \delta$, so the second contention of the theorem is satisfied if we make

$\delta \leqslant 5\lambda/6$, by making $\delta = \delta^3$, which determines $\varepsilon_2 = \varepsilon_2{}^4$ (which, as in Theorem IV.1, determines $\delta_1$, which determines $\varepsilon_1$, which determines $\varepsilon = \varepsilon^5$) and $M_1 = M_1{}^4$, as in the proof of Proposition IV.1.

Finally, we take $M_1 = \max (M_1{}^i, i = 0, 1, 2, 4)$, $M_2 = \max (M_2{}^i, i = 0, 1)$, and $\varepsilon = \min (\varepsilon^i, i = 0, 1, 2, 3, 5)$ in the linear program of Theorem III.1; the subsequent constructions do give rise to a pair $p = [x_R(.), u(.)]$ which satisfy the contentions of the present theorem. □

We shall select values of $N = N^1$ and $L = L^1$ determined according to the analysis of Theorem III.1 by the value of $\varepsilon$ just chosen; in this manner *all* of the parameters in the constructions leading to the pair $p$ are determined once the parameter $\lambda$ is chosen; then this pair $p = [x_R(.), u(.)]$ depends only on the parameter $\lambda$, and could be indexed by it. If we choose a sequence of values of $\lambda$, $\lambda = 1/j$, $j = 1, 2, \ldots$, we obtain one such pair for each value of $j$; a sequence of such pairs is thus constructed, to be denoted by $\{p^j = [x_R{}^j(.), u^j(.)], j = 1, 2, \ldots\}$. This sequence satisfies:

**Corollary** There is a sequence $\{p^j\}$ of trajectory–control pairs, which satisfies

$$\eta - \rho_2(1/j) \leqslant \mu^j(f_0) \leqslant \eta + 1/j$$

$$\|x_R{}^j(t_b) - x_b\| \leqslant 1/j$$

$$d[x_R{}^j(t), A] \leqslant 1/j, \; t \in J,$$

for $j = 1, 2, \ldots$. The measure corresponding to the pair $p^j$ is denoted by $\mu^j$.

We must clear one final point. We have shown that $\mu^j(f_0) \to \eta$. Is it possible to improve on this value by means of a sequence which satisfies the other contentions of the corollary? Suppose that there is such a sequence, $\{\nu^j\}$; then $\nu^j(f_0) \to \eta_1 < \eta$. Such a sequence must contain a weakly*-convergent subsequence, since $\nu^j(1) = \Delta t$; let $\nu^*$ be the limit of this subsequence. Then $\nu^* \in Q$, and $\nu^*(f_0) \geqslant \eta$, a contradiction since $\nu^j(f_0) \to \eta_1 < \eta$. We have therefore achieved the best possible result with a sequence satisfying the last two conditions of the corollary.

The construction of this sequence has been long and not without difficulty. The work has been worthwhile, since we have produced a solution to the modified optimal control problem defined in Chapter 1. We shall examine these matters in the next section.

## 3 Comments

In Chapter 1 we defined in some detail a modified control problem, in which the solution is not a trajectory–control pair, but a sequence of such pairs; the conditions of admissibility tend to be satisfied as the index increases, as in the corollary above. It should be emphasized that, because of the first inequality, the solution to this modified control problem is *global*, since the number $\eta$ is the infimum of the values $\mu(f_0)$ when the measure $\mu$ varies over the set of all admissible measures, that is, of those measures satisfying the infinite set of equalities (I.14). Of course, the admissible pairs of the classical formulation can be considered to be in this set, so that the classical infimum of the functional $I$ is not smaller than $\eta$, and could be larger; it could be that the pairs of the sequence $\{p^j\}$, for sufficiently large value of the index $j$, give to the functional $I$ values which are substantially *less than the classical infimum*. Whether this actually happens is an open question; we have not found examples of this situation, nor have we have able to prove that it cannot occur. After all, the pairs in the sequence do not exactly satisfy the conditions of admissibility, as the admissible classical pairs do, and it may be that, for some dynamical conditions, performance criteria, boundary values and constraint sets, the improvement that this relaxation can bring is considerable.

Similar ideas have been discussed by other authors. For instance, Warga has introduced (in WARGA [1]) the concept of 'approximate solutions', which are sequences which tend to satisfy some conditions of membership, for somewhat simpler problems than those that occupy us now. As he points out, these conditions represent restrictions which are always measured with some error, so one is inclined to accept solutions which violate these restrictions to 'an arbitrary small extent' only. And, as we mentioned in Chapter 1, numerical methods based on the classical theory do not give estimations which satisfy the restrictions exactly. It appears, then, to be a better choice to introduce the possibility of the relaxation of the constraints from the start of the development of the theory, which in some aspects can then be made powerful.

In the first four chapters of this book we have presented in detail this theory; the results in the last two chapters indicated that it is worthwhile to study the measure-theoretical optimization problem because its solution can be used to construct controls, and trajectories, with useful properties. In the next two chapters we shall present some applications of this theory, to the development of a computational method, and the study of nonlinear controllability.

**References**

Some of the material in this chapter has not been published in quite this way before. We have followed mostly RUBIO [8] and also RUBIO [5].

# 5
# Some numerical results, and a method for global minimization

## 1 A computational scheme

It would be tempting to assume that the constructions so exhaustively studied in the previous two chapters can be used without any further refinements for estimating numerically values for the trajectory and control functions appearing in the minimizing sequence $\{p^j\}$. Alas, that sequence was constructed for strictly theoretical purposes; it was our intention to show, once and for all, that the action of the optimal measure can be approximated by pairs – those in that sequence, so that the measure-theoretical optimization problem is shown to be a worthwhile one to study. Then, because of special characteristics of this problem (among others, its linearity) special methods (linear analysis) can be used to solve the problems that arise.

Linear methods – linear programming – *can* be used in the development of efficient methods for the numerical estimation of trajectory–control pairs with the good properties of the pairs in the sequence $\{p^j\}$; the linear programs and constructions developed in the past two chapters, however, need quite a few changes if they are to be used for this purpose. The main problem encountered was related to the *feasibility* of the finite-dimensional problem defined in (III.8–9); that is, whether the set of inequalities (III.8) does have a solution. (We use the customary language of finite-dimensional linear programming; for this and other topics in this field, see GASS [1].) All the examples that we attempted to solve numerically were known to be (classically) controllable; that is, the corresponding set $W$ of admissible pairs was known to be nonempty. This of course implies that the corresponding set $Q$ of measures – defined by the set of equalities (I.14) – is nonempty; further, Theorem III.1 ensures that for a number $N = N(\varepsilon)$ sufficiently large, the inequalities (III.9), and the linear programming problem described by (III.8–9), have a solution.

However, it happened that, when the functions (IV.2)

$$\theta_s(t, x, u) = t^s, \ s = 1, 2, \ldots, \tag{V.1}$$

were chosen as those whose linear combinations are dense in the set $C_1(\Omega)$, in the manner of the analysis leading to the proof of Theorem IV.1, most of the linear programs obtained in this way had no feasible solutions – that is, the system of inequalities was incompatible – for values of $N$ which were quite large, at least for the computational facilities available at our disposal. We also tried trigonometric functions in the variable $t$, without much success – the finite-dimensional linear programming problems tended to have no feasible solutions – but found that the situation improved in a marked way when piecewise continuous, pulse-like functions of the time were chosen, in a manner to be described below.

As the functions $\phi_i$ in (III.8) we chose, as in the proof of Proposition IV.1, a number, to be denoted here by $M_1'$, of monomials in $t$ and the components of $x$, ordered as in that proof; we omitted the monomials in the variable $t$ only, so the number $M_1'$ has a somewhat different meaning from the number $M_1$ used in previous chapters.

As in the analysis leading to Theorem IV.2, we chose $M_2 = 2M_{21}n$, so that the functions $\chi_h$ in (III.9) are in fact

$$\psi_j^r(t, x, u) = x_j\psi_r'(t) + g_j(t, x, u)\psi_r(t), \; r = 1, 2, \ldots, 2M_{21}, j = 1, 2, \ldots, n,$$

for $(t, x, u) \in \Omega$, with

$$\begin{aligned} \psi_r(t) &= \sin\,[2\pi r(t - t_a)/\Delta t], \; r = 1, 2, \ldots, M_{21}, \\ \psi_r(t) &= 1 - \cos[2\pi(r - M_{21})(t - t_a)/\Delta t], \; r = M_{21}+1, M_{21}+2, \ldots, 2M_{21}. \end{aligned} \tag{V.2}$$

It was necessary to add to these a number $L$ of functions of the time only, to replace the functions (V.I) which were not found suitable, as mentioned above, so that we chosen functions, to be denoted as $f_s$, $s = 1, 2, \ldots, L$, and defined as

$$\begin{aligned} f_s(t) &= 1, \quad t \in J_s \\ &= 0, \quad \text{otherwise}, \end{aligned} \tag{V.3}$$

with $J_s = (t_a + (s - 1)d, t_a + sd)$, and $d = \Delta t/L$. These functions are not continuous, and two remarks need to be made concerning their suitability:

(i) Each of the functions $f_s$, $s = 1, 2, \ldots, L$, is the limit of an increasing sequence of positive continuous functions, $\{f_{sk}\}$; then, if $\mu$ is any positive Radon measure on $\Omega$, $\mu(f_s) = \lim_{k\to\infty} \mu(f_{sk})$.

(ii) Consider now the set of all such functions, for all positive integers $L$. The linear combinations of these functions can approximate a function in $C_1(\Omega)$ arbitrarily well, in the sense that the essential supremum (see FRIEDMAN [1]) of the error function can be made to tend to zero by choosing in an appropriate manner a sufficient number of terms in the corresponding expansion.

The analysis leading to Theorem IV.1 can then be carried out in much the same way as before. It is unfortunate that in that analysis we used some of the same subindices as in the present development, for quite different purposes; we shall prime the indices of that analysis, so that $i$ becomes $i'$, etc. If the numbers $t_{i'}$, $i' = 1, 2, \ldots, R$, of the analysis leading to Theorem IV.1 are chosen to be equidistant, so that $t_{i'} - t_{i'-1} = h > 0$ for all relevant $i'$, and $h = q\Delta t/L$ for some integer $q$, then the first two terms in the definition of the number $\delta_{i'}{}^L$ are exactly zero, because the approximation is exact, the error function $F_{i'} - P_{i'}{}^L$ being zero everywhere.

When setting up the linear programming problem akin to (III.8–9), it was decided to take the parameter $\varepsilon$ as zero, at least formally; of course, the error present in the numerical computations will ensure that the solution of the linear programming problem will not satisfy exactly the constraint equations. The linear programming problem consists therefore in minimizing the linear form

$$\sum_{j=1}^{N} \alpha_j f_0(z_j) \tag{V.4}$$

over the set of coefficients $\alpha_j \geqslant 0$, $j = 1, 2, \ldots, N$, such that

$$\begin{aligned} &\sum_{j=1}^{N} \alpha_j \phi^g{}_i(z_j) = \Delta\phi_i,\ i = 1, 2, \ldots, M_1{}' \\ &\sum_{j=1}^{N} \alpha_j \chi_h(z_j) = 0,\ h = 1, 2, \ldots, M_2 \\ &\sum_{j=1}^{N} \alpha_j f_s(t_j) = a_s,\ s = 1, 2, \ldots, L; \end{aligned} \tag{V.5}$$

we have written $a_s$ for the integral of $f_s$ over $J$, the functions $\chi_h$ are those in (V.2), and $z_j$ stands, as in previous chapters, for the triple $z_j = (t_j, x_j, u_j)$. It should be remembered that the set $\omega = \{z_j, j = 1, 2, \ldots\}$ has been chosen as being dense in $\Omega$; in practice, the set $\omega^N = \{z_j, j = 1, 2, \ldots, N\} \subset \omega$ was constructed by dividing the appropriate intervals into a number of equal subintervals, defining in this way a grid of points. For instance, if $n = m = 1$, that is, we are considering a problem with one state variable and one control, and ten points are taken in each set $J$, $A$ and $U$, then $N = 1000$. The number of equations is $M = M_1{}' + M_2 + L$.

Before going into the details of the computational method used, and then giving some results, we must consider the construction of the pairs of the form $[x_R(.), u(.)]$ from the solution $\{\alpha_j, j = 1, 2, \ldots, N\}$ of the linear programming problem (V.4–5). Of course, we only need to construct the control function, since the trajectory is then simply the corresponding solution of the differential equation (I.1), with $x(t_a) = x_a$, which can be estimated numerically. The construction of the control function is of course based on the methods introduced in Chapter 4, in particular in the analysis leading to Theorem IV.1. We need to consider the followong points:

(i) Since we are taking $\varepsilon = 0$, the numbers $\delta_{i'}{}^{L}$ are all zero, since $\sigma_i(\varepsilon) \to 0$ as $\varepsilon \to 0$. We are assuming that the other two terms in the expression for $\delta_{i'}{}^{L}$ are zero, because the corresponding indices have been chosen in the manner explained above. It follows from these considerations that $H_{i'j'} = K_{i'j'}$, for all $i', j'$.

(ii) It is well known (see GASS [1]) that the linear form (V.4) attains its minimum at an extreme point of the set in $R^N$ defined by the equations (V.5) and the requirement that $\alpha_j \geqslant 0, j = 1, 2, \ldots, N$, and that such an extreme point has at most a number $M$ of nonzero elements.

(iii) Suppose that the partitions of $J$ and $A \times U$ introduced in (IV.4–5) are sufficiently fine that no set of the form $[t_{i'-1}, t_{i'}) \times V_{j'}$ contains more than one triple $z_j$ associated with an element $\alpha_j > 0$ of the extreme point described in (ii) above. Then if $[t_{i'-1}, t_{i'}) \times V_{j'}$ contains $z_k = (t_k, x_k, u_k)$ with $\alpha_k > 0$,

$$K_{i'j'} = \mu^*([t_{i'-1}, t_{i'}) \times V_{j'}) = \alpha_k, \tag{V.6}$$

so that, according to our construction following (IV.10), on $B_{i'j'}$ the control can take a value $u_k$ (and the trajectory a value $x_k$, but we are not interested in constructing it), since the pair $(x_k, u_k)$ is in $V_{j'}$.

(iv) If several triples such as $z_k$ with $\alpha_k > 0$ are contained in sets of the form $[t_{i'-1}, t_{i'}) \times V_{j'}$ with the same value of the index $i'$, then the sum of the corresponding values of $\alpha_k$ must add up to $\Delta_i$, the length of the interval $[t_{i'-1}, t_{i'})$. On each subinterval of $[t_{i-1}, t_{i'})$ the control functions takes values as described in (iii).

(v) We identify the sets of the form $[t_{i'-1}, t_{i'})$ with the sets $J_s$ defined above when introducing the functions $f_s$, in (V.3). The last set of equalities in (V.5) ensures that there will be at least one value $t_k$, a component of a triple $z_k$ associated with a component $\alpha_k > 0$ of the extreme point introduced in (ii), in each of the sets $J_s$. Note that there is no need to construct explicitly the partition of $J \times A$ used in this analysis.

(iv) We summarize the procedure for constructing the control function $u(.)$ derived from a solution of the linear programming problem (V.4–5). For $s = 1, 2, \ldots, L$, we identify the indices $k$ such that the components $\alpha_k$ of the extreme point are positive and the corresponding values $t_k$ associated with them as in v) are in $J_s$; we then partition this subinterval into further subintervals, one for each index $k$ with these properties, of length equal to the component $\alpha_k$, and make $u(t) = u_k$ on it. These subintervals which partition $J_s$ can be put together in any order.

In the next section we shall give some numerical results obtained using these techniques.

## 2 Some numerical results

We have estimated the solutions of several control problems using the techniques developed in the last chapter. Before presenting the results, it is appropriate to make several comments:

(i) The main difficulty experienced in applying this method is the very large number of unknowns in the linear programs (V.4–5); even for very elementary control problems we get values for $N$ of the order of 1000, as estimated above. Of course, the number $N$ increases geometrically with the number of dimensions of the set $\Omega$. The number of equations tends to be moderate – less than 100, even for fairly large problems, seems to be a good estimate.

(ii) The solutions of the linear programs of the form (V.4–5) were estimated by means of a method known as the *revised simplex method* (see GASS [1], Chapter 6); besides its usual attractive features, it was chosen here because it uses for all its iterations the *unmodified* matrix of coefficients. If, because of the large value of the number $N$, this matrix is too large for the available computer memory, the entries in this matrix can be computed anew every time they are needed, and only a matrix of dimensions $M \times M$ needs to be stored. In the examples presented in this chapter it was not necessary to do this, because the available memory of 4 Mbytes was (just) big enough to hold the whole of the matrix of coefficients. Of course, the time needed to solve a given problem is higher if the entries of the matrix are computed every time they are needed; we have run some tests for the purpose of comparison, and found that it increased by a factor of between 10 and 15. All the times quoted below are for programs in which the whole matrix of coefficients was stored at the beginning of the program.

(iii) The revised simplex method was home-made, rather than taken from a library of such programs, so that full access was had to every number, by-product of any computation, at any time. No attempt was made to optimize the speed of this simplex program, and it is likely that a program taken from a library would be much faster. Also, all these programs were run – in a large Amdahl computer – using a rather old-fashioned compiler; it appears that using the latest version the running times would go down by a factor of 2 to 5.

(iv) We present below the data and results obtained for eight optimal control problems; five more will be presented in the next section, in connection with a global optimization method. The following information is important:

(a) The time interval $J$ has been normalized to $J = [0, 1]$ in all cases.

(b) The sets of the form $\{z_j, j = 1, 2, \ldots, N\}$ were constructed by dividing the appropriate intervals into a number of equal sub-intervals, defining in this way a grid of points. The interval $J$ was divided into 10 intervals, the values $t_j$ were taken as 0.05, 0.15, 0.25, ..., 0.85, 0.95. The points in the $A$ set were 0, $\Delta x$, $2\Delta x$, ..., for $n = 1$, and similarly for $n = 2$ and the set $U$; the parameters $\Delta x$, $\Delta u$, etc., will be given when describing each problem, since it was found necessary to change these values according to the particular problems.

(c) The values of several parameters and of some results of the computations will be given in each case. Some of them – $M_1'$, $M_2$, $L$, $M$, $N$ – have already been defined; the following have not:

$P_1$: the number of iterations in the revised simplex method needed to obtain a basis.
$P_2$: the number of iterations in the revised simplex method needed to perform the actual optimization. The total number of iterations is then $P_1 + P_2$.
$T$: total running time.
$I^*$: optimal value of the functional $I$, as estimated by our method.
$I_c$: optimal value for the functional $I$, as estimated classically, usually by means of the principle of the Maximum. In several examples this value is not available.

**Example 1**

$\dot{x} = u \qquad x(0) = 0 \quad x(1) = 0.5 \qquad [x(1) = 0.499]$

$$I = \int_J x(t)^2 \mathrm{d}t$$

$A = [0, 1] \quad \Delta x = 0.11$
$U = [0, 1] \quad \Delta u = 0.11$
$M = 20 \quad N = 1000 \qquad I^* = 0.041758$
$M_1' = 2 \quad M_2 = 8 \quad L = 10 \qquad I_c = 0.041667 = 1/24$
$P_1 = 25 \quad P_2 = 44$
$T = 4.2$ seconds

*Comments* The trajectory and control functions are shown in Figure 2; they are very close to the classical optimal functions. The value of $x(1)$ given in square brackets is the one actually obtained, as opposed to the desired one, $x(1) = 0.5$. It should be noted that the trajectory stays for all $t \in J$ in the set $A$.

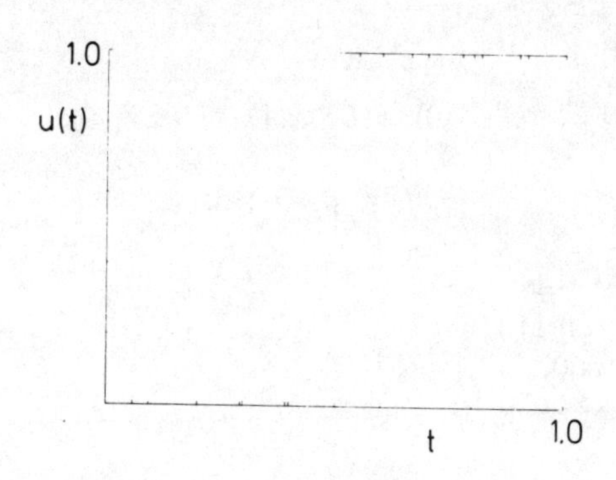


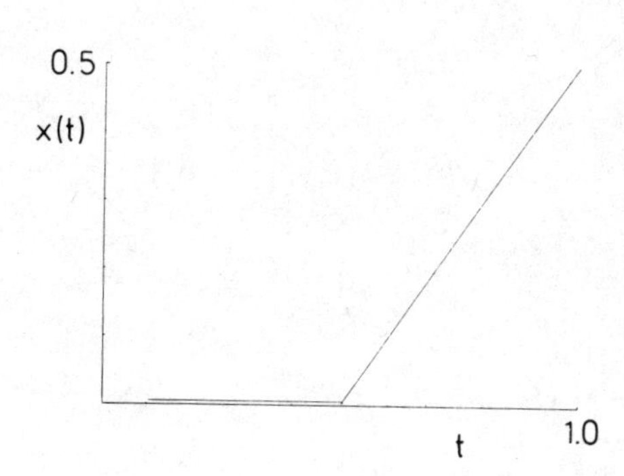


Fig. 2. Trajectory and control functions for Example 1.

**Example 2**

$\dot{x} = u$ $\qquad x(0) = 0 \quad x(1) = 0.5 \qquad [x(1) = 0.499]$

$$I = \int_J u(t)^2 \mathrm{d}t$$

$A = [0, 1] \quad \Delta x = 0.11$
$U = [0, 1] \quad \Delta u = 0.11$
$M = 20 \quad N = 1000$ $\qquad I^* = 0.25308$
$M_1' = 2 \quad M_2 = 8 \quad L = 10$ $\qquad I_c = 0.25 = 1/4$
$P_1 = 25 \quad P_2 = 44$
$T = 4.1$ seconds

*Comments* The trajectory stays in the set $A$ for all $t \in J$.

**Example 3**

$\dot{x} = (1/2)x + u \qquad x(0) = 0 \quad x(1) = 0.5 \qquad [x(1) = 0.4978]$

$$I = \int_J u(t)^2 \mathrm{d}t$$

$A = [0, 1] \quad \Delta x = 0.11$
$U = [0, 1] \quad \Delta u = 0.11$
$M = 20 \quad N = 1000 \qquad I^* = 0.1451$
$M_1' = 2 \quad M_2 = 8 \quad L = 10 \qquad I_c = 0.14549$
$P_1 = 23 \quad P_2 = 41$
$T = 2.0$ seconds

*Comments* The trajectory stays in the set $A$ for all $t \in J$.

**Example 4**

$\dot{x} = (1/2)x^2 \sin x + u \qquad x(0) = 0 \quad x(1) = 0.5 \qquad [x(1) = 0.3953]$

$$I = \int_J u(t)^2 \mathrm{d}t$$

$A = [0, 1] \quad \Delta x = 0.11$
$U = [0, 1] \quad \Delta u = 0.11$
$M = 20 \quad N = 1000 \qquad I^* = 0.1528$
$M_1' = 2 \quad M_2 = 8 \quad L = 10$
$P_1 = 23 \quad P_2 = 41$

*Comments* The desired final value is 0.5, while the one obtained is 0.3953, not nearly close enough for any foreseeable application. We must improve this by increasing the number $M_1'$. The trajectory and control functions are shown in Figure 3.

**Example 5**

$\dot{x} = (1/2)x^2 \sin x + u \qquad x(0) = 0 \quad x(1) = 0.5 \qquad [x(1) = 0.4957]$

$$I = \int_J u(t)^2 \mathrm{d}t$$

$A = [0, 1] \quad \Delta x = 0.11$
$U = [0, 1] \quad \Delta u = 0.11$
$M = 24 \quad N = 1000 \qquad I^* = 0.2425$
$M_1' = 6 \quad M_2 = 8 \quad L = 10$
$P_1 = 24 \quad P_2 = 71$
$T = 12$ seconds

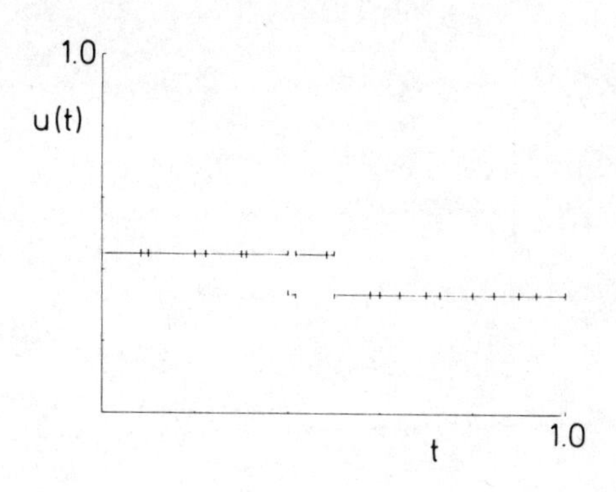


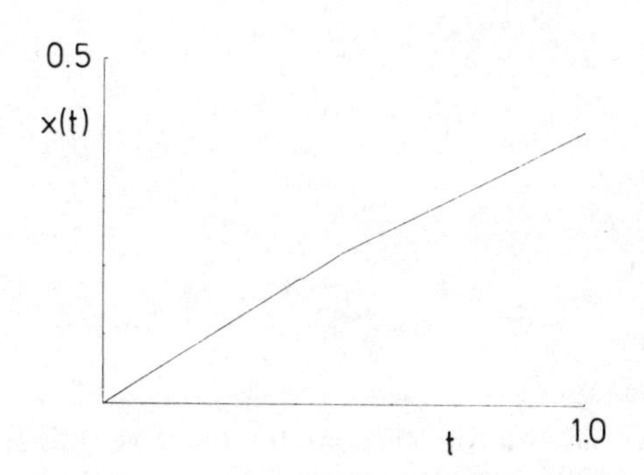


Fig. 3. Trajectory and control functions for Example 4. The final value of the trajectory is 0.3953, not 0.5 as required, and definitely not acceptable.

*Comments* The final value $x(1) = 0.4975$ is acceptable now; a (comparatively moderate) increase in the number of equations brought this improvement about, which is, of course, predicted by the theory. The trajectory and control functions are shown in Figure 4; the broken line in the graph for the trajectory shows the trajectory corresponding to Example 4.

In examples with state spaces of dimension 2, we used very coarse grids in the $A$ and $U$ sets, obtaining, however, highly accurate results:

**Example 6**

$$\dot{x}_1 = x_2 \qquad x_1(0) = 0 \quad x_1(1) = 0.1 \qquad [x_1(1) = 0.0998]$$

$$\dot{x}_2 = u \qquad x_2(0) = 0 \quad x_2(1) = 0.3 \qquad [x_2(1) = 0.2999]$$

$$I = \int_J u(t)^2 \mathrm{d}t$$

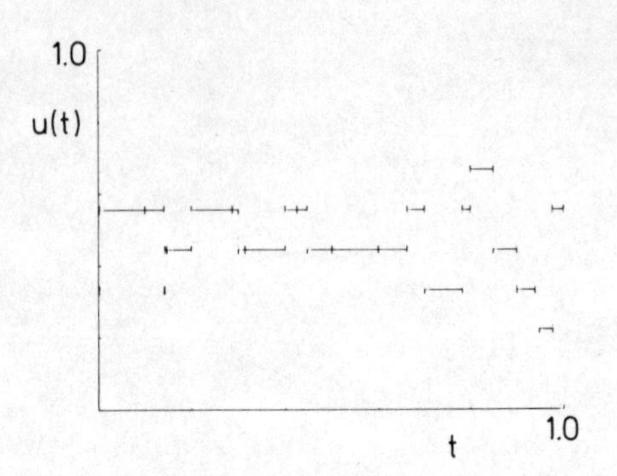


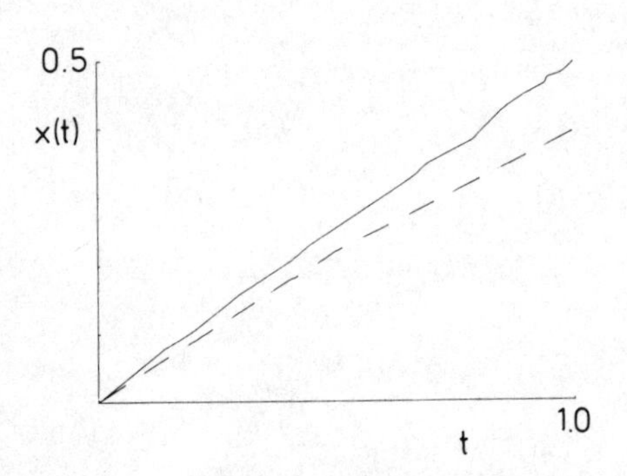


Fig. 4. Trajectory and control functions for Example 5. The final value of the trajectory is now 0.4957; this was achieved by increasing the number of relevant constraints. The broken line is a reproduction of the trajectory corresponding to Example 4.

$A = [0, 1] \times [0, 1] \quad \Delta x_1 = \Delta x_2 = 0.25$
$U = [0, 1] \quad \Delta u = 0.25$
$M = 24 \quad N = 1250$ $\qquad I^* = 0.1268$
$M_1' = 6 \quad M_2 = 8 \quad L = 10$ $\qquad I_c = 0.12$
$P_1 = 104 \quad P_2 = 141$
$T = 15.6$ seconds

*Comments* The trajectory stays in the set $A$ for all $t \in J$, as can be seen in Figure 5.

**Example 7**

$\dot{x}_1 = x_2 \qquad x_1(0) = 0 \quad x_1(1) = 0.1 \qquad [x_1(1) = 0.0997]$
$\dot{x}_2 = 10x_1{}^3 + u \qquad x^2(0) = 0 \quad x_2(1) = 0.3 \qquad [x_2(1) = 0.288]$

$$I = \int_J [x_1(t)^2 + x_2(t)^2]\mathrm{d}t$$

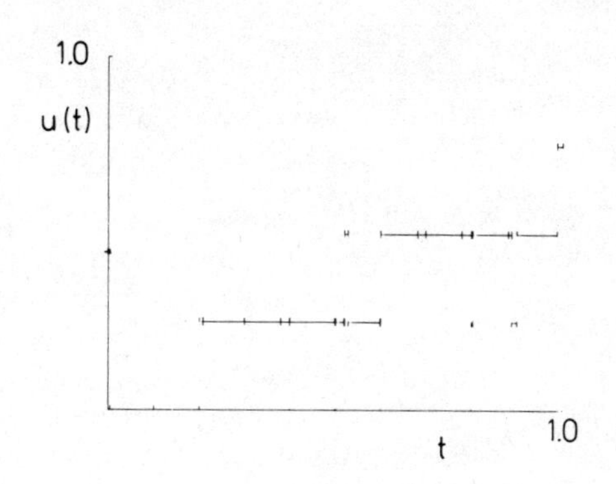


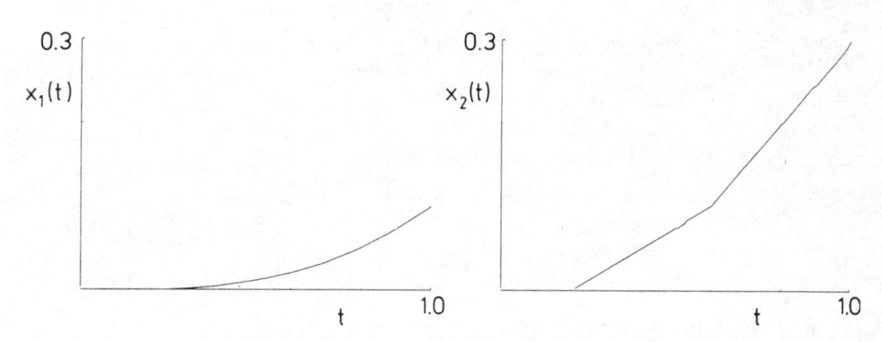


Fig. 5. Trajectory and control functions for Example 6.

$A = [0, 1] \times [0, 1] \quad \Delta x_1 = \Delta x_2 = 0.25$
$U = [0, 1] \quad \Delta u = 0.25$
$M = 24 \quad N = 1250$ $\qquad I^* = 0.024$
$M_1' = 6 \quad M_2 = 8 \quad L = 10$
$P_1 = 103 \quad P_2 = 21$
$T = 8.5$ seconds

*Comments* Again, as in all our examples, the trajectory stays in $A$ for $t \in J$. The control and trajectory functions can be seen in Figure 6. We could improve on the value of $x_2(1) = 0.288$ by increasing the number $M$.

In the next section we apply the same methods to the minimization of real-valued functions of several real variables, by transforming such problems into appropriate control problems.

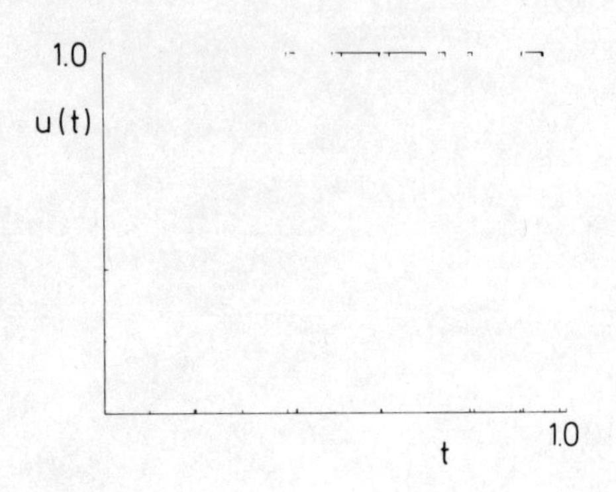


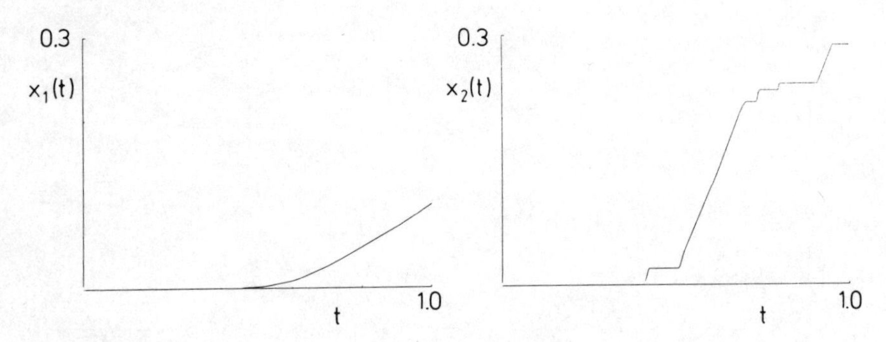


Fig. 6. Trajectory and control functions for Example 7.

### 3 A method for global optimization

Consider a control problem in which the function $f_0$ depends only on the vector $x$; the controlled system is described by the equation

$$\dot{x} = u, \tag{V.7}$$

and the control function can take only three values, $-R$, 0, $R$; that is, $u_i(t) \in \{-R, 0, R\}$, $i = 1, 2, \ldots, n$, with $R > 0$. Then the optimal trajectory will consist mostly of one point, $x(t) = x^*$, where $x^*$ is a point at which the Lipschitz function $f_0$ attains its minimum over the compact set $A \subset R^n$. There will be, of course, transients at the beginning and at the end of the interval $J$, while the initial and final points have just been left or will shortly be arrived at; these will be sufficiently short-lived if the number $R$ is sufficiently large, so that the point $x^*$ can under these conditions be identified without much difficulty. An approximation to this situation can be seen in Figure 7, corresponding to one of the problems whose solution will be given below. Note that the controls are zero for most of the time, taking the value $R$ at the beginning and at the end of the interval $J$.

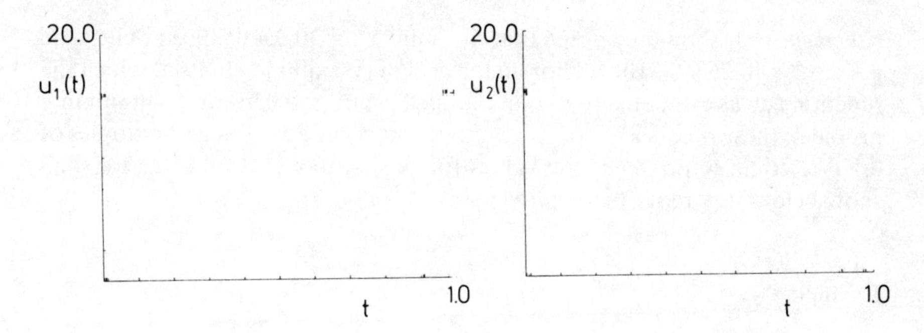


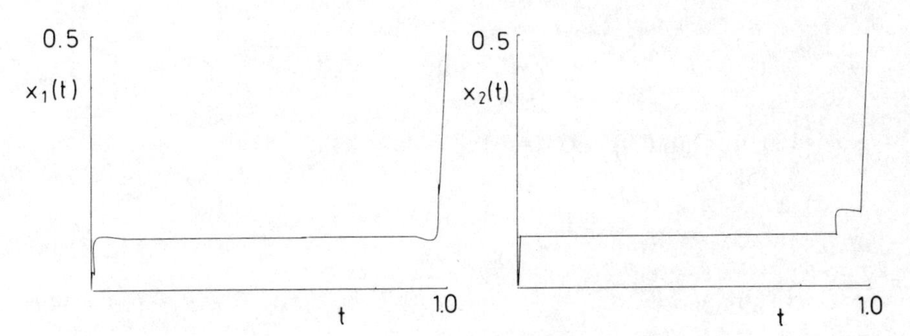


Fig. 7. Trajectory and control functions for Example 9. Some very narrow pulses can be seen around $t = 0.6$ in the graphs for the controls; these departures from the minimum do not greatly influence the steady state values, which are estimations of the coordinates of the point at which the function $f_0$ attains its minimum.

In the examples below, we have given the name $x'$ to the approximation obtained of the optimal point $x^*$.

**Example 8**

$\dot{x} = u_1$ $\qquad x_1(0) = 0 \quad x_1(1) = 0.5$

$\dot{x}_2 = u_2$ $\qquad x_2(0) = 0 \quad x_2(1) = 0.5$

$$I = \int_J [(x_1 - 0.25)^2 + (x_2 - 0.32)^2]\mathrm{d}t$$

$A = [0, 0.9] \times [0, 0.9] \quad \Delta x_1 = \Delta x_2 = 0.1$

$U = \{0, 20\} \times \{0, 20\}$

$M = 24 \quad N = 4000$

$M_1' = 6 \quad M_2 = 8 \quad L = 10$

$P_1 = 87 \quad P_2 = 227$

$T = 60.7$ seconds

$x_1' = 0.262$

$x_2' = 0.310$

$f_0(x') = 2.44 \times 10^{-4}$

$x_1^* = 0.25$

$x_2^* = 0.32$

$f_0(x^*) = 0$

*Comments* It should be noted that the point $x^* = [0.25, 0.32]$, as well as the point $x' = [0.262, 0.310]$ are not on the grid of coordinate points at which the function $f_0$ was evaluated to obtain the coefficients in the linear programming problem; these points are, for the components $x_1$ and $x_2$, integer multiples of 0.1. We could improve on the values for the components of $x'$, as we shall show below in a more demanding case.

**Example 9**

$$\dot{x}_1 = u_1 \qquad x_1(0) = 0 \quad x_1(1) = 0.5$$
$$\dot{x}_2 = u_2 \qquad x_2(0) = 0 \quad x_2(1) = 0.5$$

$$I = \int_J (\sin r/r)\mathrm{d}t, \text{ with } r = 30\sqrt{(x_1{}^2 + x_2{}^2)}$$

$A = [0, 0.9] \times [0, 0.9] \quad \Delta x_1 = \Delta x_2 = 0.1$
$U = \{0, 15\} \times \{0, 15\}$ $\qquad r' = \sqrt{(x_1{}'^2 + x_2{}'^2)} = 0.1442$
$M = 24 \quad N = 4000$ $\qquad f_0(x') = -0.2142$
$M_1{}' = 6 \quad M_2 = 8 \quad L = 10$
$P_1 = 118 \quad P_2 = 117$
$T = 47$ seconds $\qquad r^* = \sqrt{(x_1{}^{*2} + x_2{}^{*2})} = 0.1497$
$f_0(x^*) = -0.2172$

*Comments* There are some local minima of the function $f_0$ in the set $A$, for instance a whole line of them at $r = 0.3633$; however, we obtained the *global minimum* for this function of two variables, as predicted by the theory. Note, again, that the points $x'$ and $x^*$ are not on the grid. The trajectories and controls are illustrated in Figure 7. Note that this is only an approximation to the optimal situation; the controls are not zero outside neighbourhoods of the initial and final points, but exhibit some very narrow pulses around $t = 0.6$; these give rise to minute changes in the values of the trajectories. These perturbations influence only the third significant figure in the values of $x_1{}'$ and $x_2{}'$, and were taken into account by taking an average of the values of these variables before and after the pulses occurred. Similar phenomena took place in the other examples.

**Example 10**

$$\dot{x}_1 = u_1 \qquad x_1(0) = 0 \quad x_1(1) = 0.5$$
$$\dot{x}_2 = u_2 \qquad x_2(0) = 0 \quad x_2(1) = 0.5$$

$$I = \int_J [900(x_2 - 3x_1{}^2)^2 + (1 - 3x_1)^2]\mathrm{d}t$$

$A = [0, 0.9] \times [0, 0.9]$ $\Delta x_1 = \Delta x_2 = 0.1$
$U = \{0, 20\} \times \{0, 20\}$ $x_1' = 0.39$
$M = 24$ $N = 4000$ $x_2' = 0.49$
$M_1' = 6$ $M_2 = 8$ $L = 10$ $f_0(x') = 1.051$
$P_1 = 87$ $P_2 = 263$
$T = 64$ seconds $x_1^* = 1/3$
$x_2^* = 1/3$
$f_0(x^*) = 0$

*Comments* This function $f_0$ is much used to test optimization procedures because it varies very steeply near the minimum at $x^*$. Of course, our results here are not good enough; the grid is too coarse for such a rapidly varying function. However, this result gave *some* information, so that we could assume that the minimum would occur inside a new, smaller set $A = [0.2, 0.6] \times [0.2, 0.6]$, so that the grid point for the components of $x'$ are now integer multiples of 0.04:

**Example 11**

$\dot{x}_1 = u_1$ $x_1(0) = 0$ $x_1(1) = 0.5$
$\dot{x}_2 = u_2$ $x_2(0) = 0$ $x_2(1) = 0.5$

$$I = \int_J [900(x_2 - 3x_1^2)^2 + (1 - 3x_1)^2]\mathrm{d}t$$

$A = [0.2, 0.6] \times [0.2, 0.6]$ $\Delta x_1 = \Delta x_2 = 0.04$
$U = \{0, 20\} \times \{0, 20\}$ $x_1' = 0.32$
$M = 24$ $N = 4000$ $x_2' = 0.327$
$M_1' = 6$ $M_2 = 8$ $L = 10$ $f_0(x') = 0.3544$
$P_1 = 124$ $P_2 = 319$
$T = 81$ seconds $x_1^* = 1/3$
$x_2^* = 1/3$
$f_0(x^*) = 0$

*Comments* The result is much closer to the minimizing point, but the function is so rapidly varying that it takes a rather large value at $x'$; we must improve again. We can safely say that the minimum occurs in a set $A = [0.3, 0.4] \times [0.3, 0.4]$.

**Example 12**

$\dot{x}_1 = u_1$ $x_1(0) = 0$ $x_1(1) = 0.5$
$\dot{x}_2 = u_2$ $x_2(0) = 0$ $x_2(1) = 0.5$

$$I = \int_J [900(x_2 - 3x_1^2)^2 + (1 - 3x_1)^2]\mathrm{d}t$$

$A = [0.3, 0.4] \times [0.3, 0.4]$ $\Delta x_1 = \Delta x_2 = 0.01$
$U = \{0, 20\} \times \{0, 20\}$
$M = 24$ $N = 4000$
$M_1' = 6$ $M_2 = 8$ $L = 10$
$P_1 = 140$ $P_2 = 383$
$T = 103$ seconds

$x_1' = 0.3334$
$x_2' = 0.3330$
$f_0(x') = 1.96 \times 10^{-4}$

$x_1^* = 1/3$
$x_2^* = 1/3$
$f_0(x^*) = 0$

*Comments* Note that $x'$ and $x^*$ are not on the grid.

## 4 Comments

In seeking to assess the potentialities of our approach as a computational method, the increase of the number $N$ of variables with the dimensionality of the problem – the number $n + m$ of states and controls – is most important. Of course, the number $M$ of equations also increases with the number of dimensions, but not very fast, and the numbers involved are, anyway, comparatively small. The large size of the linear programming problem, due to the large size of the number of variables $N$, gives rise to two main difficulties. The first, the large amount of computer memory necessary to store the matrix of coefficients, can be taken care of by computing the entries of this matrix every time they are needed; this only exacerbates the second difficulty, increasing – by a factor of 10–15, as explained above – the time needed to actually solve the linear programming problem. Of course, an assessment can be seriously carried out only after much work in optimizing every aspect of the computation, using, for instance, a professionally written modified simplex program; we can mention here, however, an aspect that gives reason for optimism: the high accuracy that can be obtained with very coarse meshes, as in Examples 6 and 7. Also, it should be noted that there do not seem to be other methods guaranteed to work under general conditions, *even for low-order problems*; our method does not take any special notice of whether the differential equations are linear, or the performance criterion quadratic.

A possible way to proceed with the development of this method is to use in a first run of a problem a very coarse mesh – which implies a small value of $N$ – and then replace in a subsequent run the set $\Omega$ by a smaller subset containing the solution obtained in the first run; in this way we obtain high resolution with a comparatively small value of $N$. How to choose this subset in such a way as to ensure convergence is one of the several open questions concerning the material in this book. It should be noticed that we used such a method in the final optimization problem treated in Examples 10, 11 and 12.

Finally, we would like to call again the reader's attention to the fact that the minimizing points $x^*$, as well as the estimates $x'$, in the last examples are not on the grids of coordinate points built to evaluate, among other things, the function $f_0$ when constructing the performance criteria for the linear programming problems. If we were to obtain the minimizing point by simply comparing the values of $f_0$ at the grid points, a coarse grid would imply a very inaccurate answer; our method gives a very accurate answer, even with a coarse grid and a rapidly varying function $f_0$, as shown in Examples 10, 11 and 12.

# 6
# Controllability

## 1 A first approach, not completely successful

In Chapter 1 we introduced the concept of an admissible pair. The differential equation (I.1)

$$\dot{x}(t) = g[t, x(t), u(t)], \; t \in J^0 \tag{VI.1}$$

was considered there; a trajectory–control pair was said to be admissible if it satisfied this equation in the sense of Carathéodory, the control function took values in the set $U$, the trajectory function took values in the set $A$, and the boundary conditions $x(t_a) = x_a$, $x(t_b) = x_b$ were satisfied. It is not easy to determine whether the set $W$ of admissible pairs is nonempty. One is usually interested in determining the emptiness or otherwise of the set $W$ for each problem in a family, rather than for a single problem, in the following manner. Let $K$ be a closed interval in the real line not consisting of one point only, $t_a$ and $t_b$ any two points in $K$, $t_a < t_b$, $J = [t_a, t_b]$, $x_a$ and $x_b$ any two elements of $A$. We consider these points as the initial and final points of trajectories, as in our usual formulation, and they are fixed in a particular problem, even though we shall want to consider all problems as these points take values in $K \times A$, with $t_a < t_b$. If *all* the sets $W$ of admissible pairs corresponding to each of these problems are nonempty, we say that the system (VI.1), or (I.1), is *controllable*. This is not a property of the system alone; of course the sets $A$, $U$, $K$ play a fundamental role in determining whether a system is, or is not, controllable.

It is not easy to determine the controllability of nonlinear systems. It may be that the study of the corresponding concept – weak controllability – for our modified control problem is somewhat simpler, since it is possible to use the whole armoury of linear analysis.

When considering a modified control problem it is surely important to determine whether the set $Q$ of positive Radon measures associated with the problem – the set $Q$ in $\mathcal{M}^+(\Omega)$ defined by the equalities (I.14) – is nonempty. After all, if this is the case a sequence of pairs $\{[x_R^j(.), u^j(.)], j = 1, 2, \ldots\}$ can be found so that $x_R^j(t_a) = x_a$, $x_R^j(t_b) \to x_b$ and the distance $d[x_R^j(t), A] \to 0$, for $t \in J$, as $j \to \infty$, so under these conditions it is clear that, by the action of a control – $u^j(.)$, for sufficiently high $j$ – the trajectory $x_R^j(.)$

can be directed in an acceptable manner between the initial and final points. We shall say, therefore, that if $Q$ is nonempty, the points $(t_a, x_a)$ and $(t_b, x_b)$ can be *weakly joined*. If we define an interval $K$ as above, so that $J$ is a subinterval of it, we say that the system (VI.1) is *weakly controllable* if all pairs of points $(t_a, x_a)$ and $(t_b, x_b)$ in $K \times A$ for which $t_a < t_b$ can be weakly joined. Again, this property depends of course on the sets $A$, $U$, $K$.

Let us consider the initial and final points fixed. We shall establish a relationship between the emptiness or otherwise of the set $Q$ and the *problem of extensions* of linear analysis; we refer the reader to SCHAEFER [1] for the necessary background on the theory of ordered topological vector spaces, as well as for the necessary theorems on extensions. As in previous chapters, let $C(\Omega)$ be the Banach space of continuous real-valued functions on $\Omega$ with the topology of uniform convergence; the set $\Omega = J \times A \times U$ is compact since, as in the whole of this book other than in parts of Chapter 2 and in Chapter 7, the set $U$ is considered to be compact. Define the positive cone $\mathscr{C}$ of this space as the set of all functions which are nonnegative on $\Omega$; then this cone is closed in the topology of uniform convergence. Thus, $C(\Omega)$ is an *ordered topological vector space*.

Consider the subspace $E$ of $C(\Omega)$ spanned by the functions $\phi^g$, $\phi \in C'(B)$; we note that it is composed solely of functions $\phi^g$ for some $\phi \in C'(B)$. The functional $\Gamma$: $\phi^g \to \Delta\phi$ is not in general well defined, in the sense that many different functions $\phi$ can give rise to the same function $\phi^g$, so that only one value of $\Delta\phi$ cannot in general be associated with a given $\phi^g$ in $E$. Let $(t, x) \to F(t, x)$ be a continuous $n$-vector-valued function on $J \times A$, and $(t, x) \to G(t, x)$ a continuous real-valued function on $J \times A$. Suppose that

$$F(x, t)g(t, x, u) + G(x, t) = 0 \text{ on } \Omega \tag{VI.2}$$

implies that $F(x, t) = 0$, $G(x, t) = 0$, on $J \times A$. Then an equation of the type

$$F_1(x, t)g(t, x, u) + G_1(x, t) = \phi_x(x, t)g(t, x, u) + \phi_t(x, t) \text{ on } \Omega$$

has only one pair of continuous solutions, $\phi_x = F_1$, $\phi_t = G_1$ on $J \times A$; then the function $\phi$ giving rise to a given $\phi^g$ is determined up to a constant, and $\Delta\phi$ is unique; the functional $\Gamma$ is then well defined.

We assume in the argument to follow that (VI.2) does imply that $F = G = 0$ on $J \times A$. It has not been possible to find a general condition on $g$, $A$, $U$, $J$, which can guarantee this, and each special case must be studied separately. It appears that this condition is quite restrictive on the systems that can be treated in this way; thus the labelling of this as a 'not completely successful' approach, and the development of an alternative method, to be presented below. However, when it can be applied, the present method is very simple to use, and can indeed give much information, as we shall see in an example.

Suppose, then, that the functional $\Gamma$ is well defined; it maps the subspace $E$ of $C(\Omega)$ into the real line. If it is possible to extend it as a positive functional to the whole of $C(\Omega)$, then a measure exists which satisfies the equalities (I.14), and the set $Q$ is nonempty. If such a measure exists, then of course it represents a positive extension. Thus, we have

**Proposition VI.1** Let $(t_a, x_a)$ and $(t_b, x_b)$ be points in $J \times A$, with $t_a < t_b$. These points can be weakly joined if and only if the functional $\Gamma$ defined by $\Gamma(\phi^g) = \Delta\phi$, $\phi \in C'(B)$ – that is, by the equalities (I.14) on the subspace $E$ of $C(\Omega)$ – can be extended to the whole of $C(\Omega)$ as a positive functional.

There are many theorems which give necessary and sufficient conditions for such an extension to be possible; we have found the following condition, adapted from SCHAEFER [1], useful.

**Proposition VI.2** The functional $\Gamma$ defined by the equalities (I.14) on the subspace $E$ can be extended to the whole of $C(\Omega)$ as a positive functional if and only if the functional $\Gamma$ is bounded above on $E \cap (V - \mathscr{C})$, where $V$ is a suitable convex neighbourhood of zero in $C(\Omega)$.

*Proof* Let $\mu$ be a positive continuous linear extension of $\Gamma$ to $C(\Omega)$, and $V = \{F \in C(\Omega)\colon \mu(F) < 1\}$. Then $\Gamma(F) < 1$ if $F \in E \cap (V - \mathscr{C})$, so the condition is indeed necessary. Let $\Gamma$ be bounded on $E \cap (V - \mathscr{C})$, where $V$ is an open convex neighbourhood of zero in $C(\Omega)$, so that $F \in E \cap (V - \mathscr{C})$ implies that $\Gamma(F) < \alpha$ for some number $\alpha$; note that the same number $\alpha$ serves as a bound for all values $\Gamma(F)$, $F \in E \cap (V - \mathscr{C})$. Then $\alpha > 0$ and the linear manifold $S = \{F \in E\colon \Gamma(F) = \alpha\}$ does not intersect the open convex set $V - \mathscr{C}$. By the geometric form of the Hahn–Banach theorem (see TRÈVES [1], Chapter 18), there is a closed hyperplane of $C(\Omega)$, $H$, such that

$$S \subset H,\ H \cap (V - \mathscr{C}) = \varnothing;$$

thus, $H$ is of the form $H = \{F\colon \mu(F) = \alpha\}$, and $\mu$ is a continuous extension of $\Gamma$. Since $V - \mathscr{C}$ does not intersect $H$, the values $\mu(F)$, $F \in V - \mathscr{C}$, are either all higher than, or all less than, the number $\alpha$; since $0 \in V - \mathscr{C}$ and $\alpha > 0$, they must all be less than $\alpha$. Thus, since $-\mathscr{C} \subset V - \mathscr{C}$, $\mu(F) < \alpha$ for $F \in -\mathscr{C}$, or $\mu(F) > -\alpha$ for $F \in \mathscr{C}$. Also, since $F \in \mathscr{C} \Rightarrow \beta F \in \mathscr{C}$ for all $\beta > 0$, $\mu(F) > -\alpha/\beta$ for all $\beta > 0$; thus, $F \in \mathscr{C} \Rightarrow \mu(F) > 0$, and the extension $\mu$ is positive. □

A consequence of this proposition is

**Proposition VI.3** Let $(t_a, x_a)$ and $(t_b, x_b)$ be points in $J \times A$, with $t_a < t_b$. These points can be weakly joined if and only if the functional $\Gamma$ defined by $\Gamma(\phi^g) = \Delta\phi$, $\phi \in C'(B)$ – that is, by the equalities (I.14) on the subspace $E$ of $C(\Omega)$ – is bounded above on the set $\Lambda$:

$$\Lambda = \{\phi^g\colon \sup_{\Omega} \phi^g(t, x, u) < 1\} \tag{VI.3}$$

The system (VI.1) is weakly controllable if this is true for all initial and final points in $K \times A$ with $t_a < t_b$.

*Proof* The neighbourhood $V$ can be taken to be a fundamental neighbourhood, of the form

$$V = \{F\colon F \in C(\Omega), \sup_{\Omega} |F(t, x, u)| < \varepsilon\}, \varepsilon > 0.$$

Thus, the set $V - \mathscr{C} = \{F\colon F \in C(\Omega), \sup_{\Omega} F(t, x, u) < c\}$, for some positive constant $c$. According to Proposition VI.3, the extension exists if and only if the functional $\phi^g \to \Delta\phi$ is bounded above on a set of the form $\{\phi^g\colon \sup_{\Omega} \phi^g(t, x, u) < c\}$ for some positive constant $c$. This holds if and only if this functional is bounded on $\Lambda$, for which $c = 1$. Indeed, if the functional is bounded on $\Lambda$, $\sup_{\Omega} \phi^g(t, x, u) < 1 \Rightarrow \Delta\phi < \lambda$, for some constant $\lambda$. Take $c > 1$, and $\phi$ such that $\sup_{\Omega} \phi^g(t, x, u) < c$; define $\phi^g{}_1 = (1/c)\phi^g$. Then $\phi^g{}_1 \in \Lambda$, and $\Delta\phi_1 = (1/c)\Delta\phi < \lambda/c$; here $\phi_1 = (1/c)\phi \in C'(B)$. The proof in the other direction is similar. The rest of the proposition follows from Proposition VI.1. □

Of course, much hard analysis is still necessary to determine necessary and sufficient conditions in concrete cases; the conditions of Proposition VI.3 cannot be interpreted readily in terms of the properties of the differential equations and sets $A$, $U$, $J$ involved. We present such an interpretation in the next section, involving a system in which the control appears linearly.

## 2 An example

Consider the system defined by the function $g$ with components

$$g_j(t, x, u) = a_j(t, x) + \sum_{k=1}^{n} b_{jk}(t, x)u_k, \; j = 1, 2, \ldots, n, \qquad \text{(VI.4)}$$

for $(t, x, u) \in \Omega$; we assume that $U = R^n$, and that the subset $A$ of $R^n$ is bounded, closed, pathwise connected and has nonempty interior in $R^n$. An interval $K$ is chosen as in the previous section; all the coefficients in (VI.4) are assumed to be Lipschitz in $K \times A$. As above, we choose a subinterval $J$ of the interval $K$.

To study the controllability of this system, we must first determine conditions which ensure that (VI.2) implies that $F = G = 0$, which ensures that the functional $\Gamma$ is well defined. We prove:

**Proposition VI.4** The functional $\Gamma$ is well defined if the following set $\Pi$ has an empty interior in $J \times A$:

$$\Pi = \{(t, x) \in J \times A\colon \det B(t, x) = 0\}, \tag{VI.5}$$

where the $n \times n$ matrix $B(t, x) = [b_{jk}(t, x)]$.

*Proof* Write $F = (F_1, \ldots, F_n)$; then the expression (VI.2) becomes

$$\sum_{j=1}^{n} a_j(t, x)F_j(t, x) + G(t, x) + \sum_{k=1}^{n}\left[\sum_{j=1}^{n} b_{jk}(t, x)F_j(t, x)\right]u_k = 0 \tag{VI.6}$$

on $\Omega$. Fix $(t, x) \in J \times A$, at a value for which $\det B(t, x) \neq 0$. The equality (VI.6) can hold for all $u \in U = R^n$ if and only if

$$\begin{aligned} &\sum_{j=1}^{n} a_j(t, x)F_j(t,x) + G(t, x) = 0 \\ &\sum_{j=1}^{n} b_{jk}(t, x)F_j(t, x) = 0,\ k = 1, 2, \ldots, n. \end{aligned} \tag{VI.7}$$

This is system of $(n + 1)$ homogeneous equations in $(n + 1)$ unknowns, $F_j(t, x)$, $j = 1, 2, \ldots, n$, $G(t, x)$. The determinant of the system is

$$\begin{vmatrix} a_1(t, x) \ldots a_n(t, x) & 1 \\ & 0 \\ B(t, x) & \vdots \\ & 0 \end{vmatrix} = (-1)^n \det B(t, x) \neq 0;$$

then the only solution of the system (VI.7) is the zero solution, $F_j(t, x) = G(t, x) = 0$, $j = 1, 2, \ldots, n$.

Let $(t, x) \in \Pi$ now, so that $\det B(t, x) = 0$, and assume that $F_i(t, x) > 0$ for some $i$, $1 \leqslant i \leqslant n$. Then $F_i$ is positive in a neighbourhood of $(t, x)$, that is, it is positive at points of $J \times A$ which are not in $\Pi$, since this set contains no complete neighbourhood of any of its points. However $F_i$ is zero at those points of $J \times A$ which are not in $\Pi$; this contradiction shows that $F_j(t, x) = 0$ on $J \times A$, $j = 1, 2, \ldots n$. The same argument applies to the function $G$. The functional $\Gamma$ is well defined. □

If we assume that the condition of Proposition VI.4 holds, we can investigate whether the functional $\Gamma$ can be extended to the whole of $C(\Omega)$. In order to apply the results of Proposition VI.3, we must determine those functions $\phi^g$ which are in the set $\Lambda$ defined in (VI.3). Since $U = R^n$, the only functions $\phi^g$ in this set are those for which

$$\sum_{j=1}^{n} b_{jk}(t, x)\phi_{xj}(t, x) = 0;$$

this implies, by the same argument used above in the proof of Proposition VI.4, that $\phi_{xj} = 0$, $j = 1, 2, \ldots, n$; thus $\phi^g = \phi_t$ is a function of $t$ only; $\phi^g \in \Pi$ implies that $\phi_t < 1$, or $\Delta\phi < \Delta t$. Thus, the functional $\Gamma$ is bounded on $\Pi$, and any initial and final points in $K \times A$ with $t_a < t_b$ are weakly joinable. We have proved

**Proposition VI.5** Let the set $\Pi$ defined in (VI.5) have an empty interior in $K \times A$. Then the system is weakly controllable.

This proposition generalizes the well-known classical condition for the controllability of a constant system with the same number of controls and state variables. It should be noted that the result for constant systems is usually obtained by the extensive use of an explicit expression for the solution, which is not available in the more general case we have treated here.

In the next and final section of this chapter we shall develop another approach to the study of controllability, which is more general, in the sense that it can be applied to many more systems than the one we have just introduced; it is also more difficult to apply.

### 3 A more general approach

In this section we sever our connections with the optimal control and optimization problems we have studied in the rest of the book, and deal exclusively with controllability. A new framework, related to but quite different from our previous one, is to be developed; it is suitable for the study of controllability. Our previous method of studying this subject was made in the same framework in which the optimal control problems were dealt with; it was not specially designed, and was thus limited.

The limitations of this approach were due to the way in which the boundary conditions were handled – the functional $\phi^g \to \Delta\phi$ may not be well defined. In the same manner that the original equations were derived in Chapter 1, it is possible to find another way to account for the boundary conditions without introducing the functions $\phi^g$ other than in the special cases involving the equalities (I.6) and (I.9). It so happens that these equalities do not cause any trouble; as we shall see below, the corresponding functional is well defined.

Consider, then, a classical problem just like the one introduced in Chapter 1; the function $g$, sets $J$, $A$, $U$, points $(t_a, x_a)$ and $(t_b, x_b)$ play the same role as before. Let $B_a^{\delta}$ [respectively, $B_b^{\delta}$] be a closed ball in $R^{n+1}$ with centre $(t_a, x_a)$ [resp., $(t_b, x_b)$] and radius $\delta > 0$. Further, let $\Phi_a^{\delta}$ [resp., $\Phi_b^{\delta}$] be a continuous, real-valued function on $R^{n+1}$ which is positive in the interior of $B_a^{\delta}$ [resp., $B_b^{\delta}$] and zero elsewhere. Then, if $p = [x(.), u(.)]$ is an admissible pair,

$$\int_J \Phi_a^{\delta}[t, x(t)]\mathrm{d}t > 0, \quad \int_J \Phi_b^{\delta}[t, x(t)]\mathrm{d}t > 0, \qquad \text{(VI.8)}$$

since the curve $t \to x(t)$ passes trough the centre of both $B_a^{\delta}$ and $B_b^{\delta}$. We note that (VI.8) is true for any $\delta > 0$.

The equations (VI.8) play the role of the equations (I.4) in our new framework; the equations (I.6) and (I.9) are kept without change. The same process as before, in which first we consider the admissible pairs as linear positive functionals on $C(\Omega)$ which satisfy some equalities, and then redefine the control problem as one in which all such linear positive functionals satisfying these equations are considered, can be carried out in this somewhat different framework; the resulting equations are

$$\begin{aligned} \mu(\psi_j^r) &= 0, j = 1, 2, \ldots, n, r = 1, 2, \ldots, L_1, \\ \mu(f_i) &= a_i, i = 1, 2, \ldots, L_2, \\ \mu(\Phi_a^\delta) &> 0, \mu(\Phi_b^\delta) > 0, \end{aligned} \tag{VI.9}$$

where the functions $\psi_j^r$ and $f_i$ are the same as those appearing in (V.2–3). We have chosen to start with a larger set than (I.14), in which the measures satisfy a finite set of equations. The subset of $\mathcal{M}^+(\Omega)$ composed of those measures which satisfy (VI.9) will be denoted by $Q(\delta, L_1, L_2)$.

If this set is nonempty for all $L_1$ and $L_2$ and some $\delta > 0$, it is possible to develop a series of constructions such as those in Chapter 4, to generate a sequence of trajectory–control pairs with properties similar to those of the sequence $\{[x_R^j(.), u^j(.)]\}$ in Chapter 4. We say that the points $(t_a, x_a)$ and $(t_b, x_b)$ can be $(\delta, L_1, L_2)$-*weakly joined* if the set $Q(\delta, L_1, L_2)$ is nonempty.

Consider the subspace $E$ of $C(\Omega)$ spanned by the set

$$\{\psi_j^r, j = 1, 2, \ldots, n, r = 1, 2, \ldots, L_1, f_i, i = 1, 2, \ldots, L_2, \Phi_a^\delta, \Phi_b^\delta\}.$$

A functional $\Gamma$ can be defined on $E$ as follows. Let

$$\begin{aligned} \Gamma(\psi_j^r) &= 0, j = 1, 2, \ldots, n, r = 1, 2, \ldots, L_1, \\ \Gamma(f_i) &= a_i, i = 1, 2, \ldots, L_2, \\ \Gamma(\Phi_\alpha^\delta) &= S_a > 0, \\ \Gamma(\Phi_\beta^\delta) &= S_b > 0; \end{aligned} \tag{VI.10}$$

$\Gamma$ is then extended to the whole of $E$ by linearity. We have chosen fixed values for $\Gamma(\Phi_a^\delta)$ and $\Gamma(\Phi_b^\delta)$. We note that this functional is well defined on $E$, since the basis functions for this subspace are independent; of course, these functions are considered to be defined on $\Omega$, even if they depend only on $t$ or on $(t, x)$.

As in the previous approach, the problem of the existence of a measure in $Q(\delta, L_1, L_2)$ is equivalent to the problem of extending $\Gamma$ from the subspace $E$ to the whole of $C(\Omega)$ as a positive functional on this space. Propositions (VI.1–3) can be trivially adapted to the present framework; they give rise to

**Proposition VI.6** The initial and final points can be $(\delta, L_1, L_2)$-weakly joined if and only if the condition

$$\alpha\Phi_a^\delta + \beta\Phi_b^\delta + \sum_{i=1}^{L_2} \theta_i f_i < 1 - \sum_{j=1}^{n}\sum_{r=1}^{L_1} \gamma_{jr}\psi_j^r \tag{VI.11}$$

on $\Omega$, implies that there is a constant $D$ such that

$$\Gamma(F) = \alpha S_a + \beta S_b + \sum_{i=1}^{L_2} \theta_i a_i < D \qquad \text{(VI.12)}$$

for all sets $\{\alpha, \beta, \theta_i, i = 1, 2, \ldots, L_2\}$ satisfying (VI.11).

It is necessary to translate the condition in this proposition in terms of conditions on the functions and sets which form the data of the problem. We need the following definition:

*Definition* Let $\{\chi_k, k = 1, 2, \ldots, R\}$ be a set of real-valued functions on a set $Z$. This set of functions is said to have *the property S in Z* if the functions $\chi_k$ have the following property: there are $2^R$ values of $z$, $\{z_i, i = 1, 2, \ldots, 2^R\}$, such that the functions in the set take, at these values, the signs indicated in the following array:

| | $\chi_1$ | $\chi_2$ | $\ldots$ | $\chi_R$ |
|---|---|---|---|---|
| 1 | + | + | ... | + |
| 2 | − | + | ... | + |
| 3 | − | − | ... | + |
| ⋮ | | | | |
| $2^R$ | − | − | ... | − |

(VI.13)

The signs + and − indicate nonnegative and nonpositive values respectively. For instance, the second row in the array says that

$$\chi_1(z_2) \leqslant 0 \quad \chi_2(z_2) \geqslant 0, \ldots, \chi_R(z_2) \geqslant 0.$$

The array indicates that, at each value $z_i$, some of the functions are nonnegative, and the rest nonpositive; all possible combinations occur.

**Proposition VI.7** Let a variable $y$ satisfy the set of inequalities

$$y < 1 - \sum_{j=1}^{R} \xi_j \chi_j(z), \; z \in Z,$$

where the $\xi_j$'s are real numbers which can take any values. Then, $y$ is bounded above (for all values of the $\xi_j$'s) if and only if the set $\{\chi_j, j = 1, 2, \ldots, R\}$ has the property $S$ in $Z$.

*Proof* The variable $y$ can become unbounded when some $\xi_j$'s tend to $+\infty$ or to $-\infty$. Suppose, for instance, that

$$\xi_1 \to +\infty, \; \xi_4 \to -\infty,$$

while the other variables are zero. There are several values $z_i$ such that

$$\chi_1(z_i) > 0, \; \chi_4(z_i) \leqslant 0,$$

so that

$$1 - \xi_1\chi_1(z_i) - \xi_4\chi_4(z_i) \leqslant 1,$$

for *all* $\xi_1$, $\xi_4$, so that $y$ does not fail to stay bounded as these variables tend to $+\infty$ and $-\infty$ respectively. There are values of $z_i \in Z$ to take care of all possible combinations of behaviour, as some or all of the $\xi_j$'s tend to $+\infty$ or to $-\infty$. If these values are not present, $y$ will not be bounded. □

We can now apply these results to the controllability problem. Consider the division of the interval $J$ introduced when defining the functions $f_i$. Let $t_i$ be a value of $t$ in the interval

$$T_i \equiv (t_a + (i-1)d,\ t_a + ih],\ i = 2, 3, \ldots, L_2 - 1,\ d = \Delta t/L_2;$$

also, choose $t_1$, $t_{L2}$, $\tau_a$, $\tau_b$, as follows:

$$t_1 \in (t_a + \delta,\ t_a + d] \equiv T_1,\ t_{L2} \in (t_a + (L_2 - 1)d,\ t_b - \delta] \equiv T_{L2};$$

$$\tau_a \in [t_a,\ t_a + \delta) \equiv T_a,\ \tau_b \in (t_b - \delta,\ t_b] \equiv T_b.$$

We assume that $\delta < d/2$. Let $I_a{}^\delta(t_0)$ and $I_b{}^\delta(t_0)$ be the intersections of $B_a{}^\delta$ and $B_b{}^\delta$ with the plane $t = t_0$, for any $t_0 \in J$. In what follows, it will be necessary to combine many inequalities derived from the expression (VI.11); it becomes necessary therefore to distinguish between the triples $(x, u) \in A \times U$ associated with the different values of the time variable. Thus, the triples in $T_i \times A \times U \equiv H_i$ will be referred to as $(t_i, x^i, u^i)$, $i = 1, 2, \ldots, N$, and those in

$$\underset{\tau_a \in T_a}{U} \{\tau_a\} \times I_a{}^\delta(\tau_a) \times U \equiv H_a$$

as $(\tau_a, x^a, u^a)$; in a similar way we define $H_b$, its elements are those of the form $(\tau_b, x^b, u^b)$.

If (VI.11) is satisfied, we shall have, taking into consideration the definition of the functions involved,

$$\theta_i < 1 - \sum_{j=1}^{n} \sum_{r=1}^{L_1} \gamma_{jr} \psi_j{}^r(t_i, x^i, u^i) \text{ on } H_i,\ i = 1, 2, \ldots, L_2, \tag{VI.14}$$

$$\alpha\Phi_a{}^\delta(t_a) + \theta_1 < 1 - \sum_{j=1}^{n} \sum_{r=1}^{L_1} \gamma_{jr} \psi_j{}^r(\tau_a, x^a, u^a) \text{ on } H_a, \tag{VI.15}$$

$$\beta\Phi_b{}^\delta(t_b) + \theta_{L2} < 1 - \sum_{j=1}^{n} \sum_{r=1}^{L_1} \gamma_{jr} \psi_j{}^r(\tau_b, x^b, u^b) \text{ on } H_b. \tag{VI.16}$$

We can now estimate $\Gamma(F)$, as in (VI.12); beforehand, we must select values for $S_a$ and $S_b$. Since these values must tend to zero as $L_2$ tends to infinity, we have chosen $S_a = S_b = \Delta t/L_2$. Also, we have chosen the values of $\Phi_a{}^\delta$ and $\Phi_b{}^\delta$ as unity at the appropriate values of the arguments. Then,

$$\Gamma(F) = \alpha S_a + \beta S_b + \sum_{i=1}^{L_2} \theta_i a_i < 1 - \sum_{j=1}^{n} \sum_{r=1}^{L_1} \gamma_{jr} \chi_j^r(z),\ z \in Z, \quad \text{(VI.17)}$$

with

$$z \equiv [(\tau_a, x^a, u^a), (\tau_b, x^b, u^b), (t_2, x^2, u^2), \ldots, (t_{L2-1}, x^{L2-1}, u^{L2-1})],$$

$$Z \equiv H_a \times H_b \times H_2 \times \ldots \times H_{L2-1},$$

and

$$\chi_j^r(z) = (\Delta t/L_2)\left[\sum_{i=2}^{L_2-1} \psi_j^r(t_i, x^i, u^i) + \psi_j^r(\tau_a, x^a, u^a) + \psi_j^r(\tau_b, x^b, u^b)\right]$$

Considering our results and definitions above, we have proved:

**Theorem VI.1** The initial and final points can be $(\delta, L_1, L_2)$-weakly joined if and only if the set $\{\chi_j^r, j = 1, 2, \ldots, n, r = 1, 2, \ldots, L_1)$ has the property $S$ on $Z$.

As in the previous section, it is possible to define the concept of weak controllability; if $t_a$ and $t_b$ are constrained to be in an interval $K$, $t_a < t_b$, and if all pairs of points $(t_a, x_a)$, $(t_b, x_b)$ in $K \times A$ can be $(\delta, L_1, L_2)$-weakly joined, then the system is said to be $(\delta, L_1, L_2)$-weakly controllable. If a system has this property for all $\delta > 0$, $L_1$ and $L_2$, we can say that the system is *weakly controllable*. Then, we have

**Corollary** A system is weakly controllable if and only if the corresponding properties in Theorem VI.1 are satisfied for all $J \subset K$, all $\delta > 0$, all $L_1$ and $L_2$.

The development of these ideas into a practical method for the study of nonlinear controllability is an active part of our present research activity; the results will be published in the control literature.

## References

In the first two sections of this chapter we have followed RUBIO [9]; the last section is inspired on the treatment by RUBIO [10], but is a considerable improvement over the material in that paper.

# 7
# Hilbert space

## 1 Introduction

In several previous publications on these matters (see RUBIO [5–8]), we have developed the theory for systems somewhat more general than those treated in this book, for which the state space is not the euclidean space $R^n$, but a Hilbert space. The results obtained for these systems are much the same as those presented in the previous chapters; we shall examine these matters now, as well as give one further result concerning the existence of an optimal measure.

The basic elements of an optimal control problem involving a Hilbert state space are the same as in the euclidean case; we follow RUBIO [5]. Let $X$ be a separable real Hilbert space, $A$ be a bounded pathwise connected subset of $X$, $J = [t_a, t_b]$, $x_a$ and $x_b$ be two elements of $A$, and $U$ a subset of a normed linear space $Y$. Consider the strong differential equation in $X$:

$$\dot{x}(t) = g[t, x(t), u(t)] \text{ a.e. on } J, \qquad \text{(VII.1)}$$

where $\dot{x}(.)$ is the strong derivative of $x(.)$, $u(t) \in U$, $t \in J$, $u(.)$ is measurable, $g$ is a continuous *bounded* function mapping $\Omega = J \times A \times U$ into $X$, $A$ having the topology induced by $X$, and $\Omega$ the product topology. We say that a pair $p = [x(.), u(.)]$ is admissible if

(i) The functions $x(.)$ and $u(.)$ satisfy (VII.1), in the sense that

$$x(t) = x(t_a) + \int_{t_a}^{t} g[\tau, x(\tau), u(\tau)]\mathrm{d}\tau, \; t \in J;$$

the integral is a Bochner integral with respect to the Lebesgue measure on $J$. Note that this requirement implies that $x(.)$ admits a strong derivative equal to $g[t, x(t), u(t)]$ a.e. on $J$.

(ii) The boundary conditions $x(t_a) = x_a$, $x(t_b) = x_b$ are satisfied.

(iii) The trajectory function $x(.)$ satisfies: $x(t) \in A$, $t \in J$.

It is desired to minimize a functional on the set of admissible pairs:

$$I(p) = \int_J f_0[t, x(t), u(t)]\mathrm{d}t, \qquad \text{(VII.2)}$$

where $f_0$ is a *bounded* continuous function on $\Omega$. It is possible to weaken some of these conditions, especially the requirement that $g$ and $f$ are bounded functions.

The abstraction of some characteristics of the admissible pairs can now proceed much as in Chapter 1; there are, however, several differences:

(i) Let $B$ be an open ball in $R \times X$ containing $J \times A$, and $C_b{}^1$ the space of strongly continuously differentiable functions $\phi$ on $B$ such that $\phi$ and its derivatives are bounded on $B$, in the sense that $|\phi(t, x)| \leqslant M$, $\|\phi_x(t, x)\| \leqslant M$, $|\phi_t(t, x)| \leqslant M$ for $(t, x) \in B$ and some constant $M$ depending on the function $\phi$ itself. Here $\|\phi_x(t, x)\|$ is the norm of the linear functional $\phi_x(t, x)$: $X \to R$. Consider the function $\phi^g$: $\Omega \to R$ defined by

$$\phi^g(t, x, u) = [\phi_x(t, x) \circ g](t, x, u) + \phi_t(t, x).$$

For any admissible pair $p$ and all $\phi \in C_b{}^1(B)$,

$$\int_J \phi^g[t, x(t), u(t)]\mathrm{d}t = \Delta\phi.$$

(ii) Since $X$ is separable, there exists a countable set dense in $X$, and we can find a set $W = \{w_j, j = 1, 2, \ldots\}$, $w_j \in X$, such that for all $m$ the set $\{w_1, \ldots, w_m\}$ is linearly independent, and the finite linear combinations of elements of $W$ are dense in $X$. Put

$$\psi_j(t, x, u) = (x, w_j)\psi'(t) + (g(t, x, u), w_j)\psi(t),$$

$(t, x, u) \in \Omega$, with $(,)$ the inner product in $X$. Let $\psi \in \mathscr{D}(J^0)$. If $p = [x(.), u(.)]$ is an admissible pair, then for all $\psi \in \mathscr{D}(J^0)$ and $j = 1, 2, \ldots$,

$$\int_J \psi_j[t, x(t), u(t)]\mathrm{d}t = 0.$$

The third set of equalities, those for functions which depend on the time only, are the same as (I.9). Define now $C_b(\Omega)$ as the linear space of all continuous bounded real-valued functions on $\Omega$. The functions $f_0, \phi^g, \psi_j$, are in $C_b(\Omega)$, for all $\phi \in C_b{}^1(B)$, all $\psi \in \mathscr{D}(J^0)$, $j = 1, 2, \ldots$, as well as the functions $f$ which depend on the variable $t$ only appearing in (I.9). We norm this space by means of the sup norm, which makes it a Banach space.

An admissible pair $p$ can be identified with a positive linear continuous functional $\Lambda_p$ on $C_b(\Omega)$:

$$\Lambda_p: F \in C_b(\Omega) \to \int_J F[t, x(t), u(t)]\mathrm{d}t.$$

An admissible pair satisfies equalities similar to (I.11), with $C_b{}^1(\Omega)$ taking the role of $C'(B)$, and $C_{b1}(\Omega)$, the subspace of $C_b(\Omega)$ consisting of all functions in this space which depend only on the variable $t$, that of $C_1(\Omega)$.

In the case treated in the rest of the book, in which the set $\Omega$ is compact, there are powerful representation theorems for positive linear functionals such as $\Lambda_p$ in terms of Radon measures. In the present case, in which $\Omega$ is not locally compact, one has to be content with the results given, for instance, in VARADARAJAN [1], in which it is shown that to each positive continuous linear functional $\Lambda$ on $C_b(\Omega)$ one can associate a *finitely additive* positive Borel measure on $\Omega$ such that $\mu(F) = \Lambda(F)$ for all $F \in C_b(\Omega)$, and $\|\Lambda\| = \mu(1)$, with 1 the function equal to one on $\Omega$. As in the locally compact case, we modify our control problem, and are led to consider the problem of minimizing the linear function $\mu \to \mu(f_0)$ over the set $Q$ of positive, finitely additive Borel measures on $\Omega$ defined by the equalities

$$\mu(\phi^g) = \Delta\phi, \ \phi \in C_b{}^1(B),$$

$$\mu(\psi_j) = 0, \ \psi \in \mathscr{D}(J^0), j = 1, 2, \ldots,$$

$$\mu(f) = a_f, f \in C_{b1}(\Omega).$$

## 2 Existence and approximation

The proof of the existence of an optimal measure $\mu^* \in Q$ proceeds under the present hypothesis much as the proof of Theorem II.1; it is all a matter of a continuous functional defined on a compact set. It should be noted that this result has been established under the assumption that the functions involved are continuous and bounded. The most important realization of this condition occurs when these functions are continuous and map bounded sets into bounded sets, while the sets involved – $J$, $A$, $U$ – are bounded; in particular, then, the admissible controls should be bounded. We have also obtained existence results for the case in which the controls take values in an unbounded set, while the trajectories take values in a bounded set; the functions involved are continuous and map bounded sets into bounded sets. The reader is referred to RUBIO [6] for the details, which are very similar to those in the proof of Theorem II.2; in particular, the conditions (II.7–9), suitably interpreted, ensure that an optimal measure exists in $Q$.

As in RUBIO [7], we shall now assume that $f_0$ is continuous and nonnegative, rather than continuous and bounded on bounded sets. The proper generalization of the situation studied in RUBIO [6] is the one in which $f_0$ is continuous and bounded below; this situation can be reduced to the one to be considered now by the addition to $f_0$ of a suitably chosen constant.

Let $A'$, $U'$ be the Stone–Čech compactifications of $A$ and $U$ respectively (see SCHUBERT [1], Section 9, for an introduction to these matters). Then the space $\Omega' = J \times A' \times U'$, with the product topology, has the following characteristics: (i) it is compact; (ii) $\Omega$ is dense in $\Omega'$; (iii) each $F \in C_b(\Omega)$ has a unique extension $F' \in C(\Omega')$. Given $F' \in C(\Omega')$, we shall call $F$ its restriction to $\Omega$; of course, $F \in C_b(\Omega)$.

Every functional $\Lambda_p$ corresponding – as above – to an admissible pair $p$ induces a linear positive functional $\Lambda_p'$ on $C(\Omega')$:

$$\Lambda_p'(F') = \Lambda_p(F),\ F' \in C(\Omega');$$

according to Riesz's theorem, there is a Borel measure $\mu_p$ on $\Omega'$ such that

$$\Lambda_p'(F') = \mu_p(F),\ F' \in C(\Omega').$$

Of course, these measures $\mu_p$ satisfy the following equalities:

$$\begin{aligned} \mu_p(\phi^{g'}) &= \Delta\phi,\ \phi \in C_b{}^1(B), \\ \mu_p(\psi_j') &= 0,\ \psi \in \mathscr{D}(J^0), j = 1, 2, \ldots, \\ \mu_p(f') &= a_{f'}, f \in C_{b1}(\Omega). \end{aligned} \tag{VII.3}$$

The role of the function $f_0$ is somewhat different, since it is not, in general, in $C_b(\Omega)$. Its extension to $\Omega', f_0'$, is not in $C(\Omega')$ but in $C(\Omega'; \bar{R})$, the space of continuous functions on $\Omega'$ with values in $\bar{R}$, the extended real line. However, the integral

$$\mu_p(f_0') = \int_{\Omega'} f_0' \mathrm{d}\mu_p \tag{VII.4}$$

makes sense, as we shall show below. Thus, the original control problem has generated another control problem in a somewhat different setting: among all Borel measures $\mu_p$ on $\Omega'$ corresponding to admissible pairs find one which minimizes the functional defined by (VII.4); of course, these measures satisfy the conditions (VII.3). This problem, of course, does not have a solution if the original problem does not have one, and, by a now familiar argument, we are led to consider the problem of minimizing the function $\mu \to \mu(f_0')$ over all positive Borel measures $\mu$ on $\Omega'$ satisfying the conditions (VII.3). We shall show below that this problem has a solution.

Let $\mathscr{M}(\Omega')$ be the space of signed Borel measures on $\Omega'$. We put on this space the weak*-topology $\sigma(\mathscr{M}(\Omega'), C(\Omega'))$. We prove

**Proposition VII.1** The set $Q$ of positive Borel measures on $\Omega'$ defined by the equalities (VII.5) is compact, and the function $\mu \to \mu(f_0)$ is lower semi-continuous on $\mathscr{M}^+(\Omega')$; it attains its greatest lower bound on the set $Q$.

*Proof* The weak*-compactness of the set $Q$ is proved in much the same way as that of the set $Q$ of Proposition II.2. Let $f_{0n}' = \inf(f_0', n)$, where $n$ is the function equal to $n$ on $\Omega'$. Then $f_{0n}' \in C(\Omega')$, and $f_{0n}' \leqslant f_{0.n+1}'$; that is, the family $\{f_{0n}'\}$ is increasing. Then, by the monotone convergence theorem, since $f_o' \geqslant 0$ and $f_{0n}' \geqslant 0$, for any $\mu \in \mathscr{M}^+(\Omega')$,

$$\mu(f_0') = \lim_{n\to\infty} \mu(f_{0n}') = \sup_n \mu(f_{0n}'),$$

The functions $\mu \to \mu(f_{0n}')$ are continuous on $\mathscr{M}^+(\Omega')$, and increasing, since the measures are positive and the family $\{f_{0n}'\}$ is increasing. Thus, the function $\mu \to \mu(f_0)$ is the pointwise limit of a sequence of increasing continuous functions and is therefore lower semicontinuous. The final contention of the proposition follows from this fact, the compactness of $Q$ and Proposition II.1. □

It is necessary to go back now to the usual spaces, away from the Stone–Čech compactifications and their associated spaces of measures and functions. Consider the space $\mathscr{M}_L^+(\Omega')$, consisting of those measures in $\mathscr{M}^+(\Omega)$ for which $\mu(f') = a_{f'}$ for all those $f' \in C(\Omega')$ which depend only on the variable $t$; $a_{f'}$ is the Lebesgue integral of $f'$ over the interval $J$.

Since $\mu \to \mu(f_0')$ is lower semicontinuous on $\mathscr{M}^+(\Omega')$, it is also lower semicontinuous on $\mathscr{M}_L^+(\Omega')$. Then, if $\mu^*$ is the optimal measure of Proposition VII.1 – one at which the minimum of $\mu \to \mu(f_0)$ is attained over the set $Q$ – since $\mu \in \mathscr{M}_L^+(\Omega')$ we have (see CHOQUET [2], p. 132),

$$\mu^*(f_0') = \liminf_{\substack{\mu\to\mu^* \\ \mu\in\mathscr{M}_L^+(\Omega')}} \mu(f_0'); \tag{VII.5}$$

thus, $\mu^*(f_0')$ is the smallest cluster point of $\mu \to \mu(f_0')$ with respect to the filter $\mathscr{A}$ determined by the filter basis $\mathscr{B}$ consisting of all sets of the form $U = V \cap \mathscr{M}_L^+(\Omega')$, with $V$ a neighbourhood of $\mu^*$ in the weak*-topology. (See BOURBAKI [1], p. 354).

We consider now the space $\mathscr{M}_1^+(\Omega')$, consisting of those measures on $\Omega'$ representing a functional $\Lambda_p'$ corresponding to a pair $p = [x_c(.), u(.)]$, where $t \to x_c(t) \in A$, $t \in J$, is continuous and $t \to u(t) \in U$, $t \in J$, is piecewise constant. It was proved in RUBIO [5], Proposition 2, that $\mathscr{M}_1^+(\Omega')$ is weakly*-dense in $\mathscr{M}_L^+(\Omega')$; the proof is very similar to our proof of Theorem IV.1, being partly based on the work of GHOUILA-HOURI [1]. We prove that $\mu^*(f_0)$ is a cluster point of $\mu \to \mu(f_0)$ with respect to the filter $\mathscr{A}_1$ determined by the filter basis $\mathscr{B}_1$ consisting of all sets of the form $U = V \cap \mathscr{M}_1^+(\Omega')$, with $V$ as above. Indeed, $\mu \to \mu(f_{0n}')$ has a limit point at $\mu^*(f_{0n}')$ with respect to the filter $\mathscr{A}_1$, since this function is continuous at $\mu^*$ and $\mathscr{M}_1^+(\Omega')$ is dense in $\mathscr{M}_L^+(\Omega')$; this implies that there exists a set $U_n \in \mathscr{A}_1$ such that

$$|(\mu - \mu^*)f_{0n}'| < 1/n \text{ for all } \mu \in U_n.$$

Take any measure $\mu_n \in U_n$. Then

$$|(\mu^* - \mu_n)f_0'| \to 0 \text{ as } n \to \infty,$$

since

$$|(\mu^* - \mu_n)f_0'| \leqslant |(\mu^*(f_0' - f_{0n}')| + |(\mu^* - \mu_n)f_{0n}'| + |(\mu_n(f_{0n}' - f_0')|.$$

Therefore, $\mu^*(f_0')$ is a cluster point of $\mu \to \mu(f_0')$ with respect to $\mathscr{A}_1$. There is, therefore, a net $\{\mu_{p\alpha}\}$ of measures $\mu_{p\alpha}$ defined by pairs $p_\alpha = [x_{c\alpha}(.), u_\alpha(.)]$ converging weakly* to $\mu$, such that

$$\mu_{p\alpha}(f_0') = \mu_{p\alpha}(f_0) \to \mu^*(f_0') \leqslant \mu_p(f_0') = \Lambda_p(f_0),$$

for all admissible pairs $p$. By means of members of the net $\{\mu_{p\alpha}\}$ we can achieve, therefore, values for the performance functional which are at least close to the greatest lower bound of the values achieved on the set of admissible pairs; the situation here is similar to that prevailing in the locally compact case. Also, since the measure $\mu^*$ satisfies the equalities (VII.3), and the net $\{\mu_{p\alpha}\}$ converges weakly* to $\mu^*$,

$$\mu_{p\alpha}(\phi^g) \to \Delta\phi,\ \phi \in C_b{}^1(B),$$

$$\mu_{p\alpha}(\psi_j) \to 0,\ \psi \in \mathscr{D}(J^0),\ j = 1, 2, \ldots$$

Of course, the measures in the net $\{\mu_{p\alpha}\}$ are in $\mathscr{M}_L{}^+(\Omega')$, and satisfy automatically the last set of equalities in (VII.3).

It is possible now to apply the approximation techniques of Chapters 3 and 4 to this case, with only minor variations; it is necessary to postulate the existence of functions in $C_b{}^1(B)$ which play the role of the monomials $\phi_i$, $i = 1, 2, \ldots$, which appear in Chapter 3. We refer the reader to RUBIO [8] for the – fairly routine – details.

# 8
# The diffusion equation

## 1 Existence

We consider in this chapter the one-dimensional diffusion equation

$$y_{xx}(x, t) = y_t(x, t), (x, t) \in (0, 1) \times (0, T), \qquad \text{(VIII.1)}$$

with boundary conditions

$$\begin{aligned} y_x(0, t) &= 0 \qquad t \in [0, T], \\ y_x(1, t) &= u(t) \quad\; t \in [0, T], \\ y(x, 0) &= 0 \qquad x \in [0, 1], \end{aligned} \qquad \text{(VII.2)}$$

where $t \in [0, T] \to u(t) \in R$ is the control function. We consider a classical control problem associated with this equation and its boundary conditions; this problem can be treated by means akin to those employed in the previous chapters to deal with systems described by ordinary differential equations.

The control function $u(.)$ will be termed *admissible* if it is a measurable function on $[0, T]$ and

(i) It takes values in the set $[-1, 1]$ for $t \in [0, T]$.
(ii) The solution of the partial differential equation (VIII.1) corresponding to the boundary conditions (VIII.2) satisfies a terminal condition,

$$y(T, x) = g(x) \quad \text{Lebesgue-a.e. for } x \in [0, 1], \qquad \text{(VIII.3)}$$

where $g(.) \in L_2(0, 1)$ is the desired final state. The set of all admissible controls will be designated by $V$; it is assumed nonempty. This is, of course, an assumption of *controllability*: it is possible to reach the state $g(.)$ at time $T$ from the state $y(x, 0) = 0$, $x \in [0, 1]$, by means of an admissible control. We shall examine this assumption in some detail in Section 2 of the present chapter.

The classical control problem consists of finding an admissible control $u(.)$ which minimizes the functional

$$I[u(.)] = \int_0^T f_0[t, u(t)]\mathrm{d}t, \qquad \text{(VIII.4)}$$

where $f_0 \in C(\Omega)$, the space of continuous functions on $\Omega = [0, T] \times [-1, 1]$ with the topology of uniform convergence.

This control problem may or may not have a solution in $V$. The same philosophy guiding the rest of the book indicates that this set should be somewhat enlarged, to a set of measures into which the set $V$ can be injected. This can be done in the present case by restating the problem in terms of a moment problem.

We consider the solution of (VIII.1) and (VIII.2) in the sense defined by FATTORINI and RUSSELL [1], in which case

$$y(x, T) = \int_0^T u(t)\mathrm{d}t + \sum_{n=1}^{\infty} 2(-1)^n \int_0^T \exp\,[-n^2\pi^2(T - t)]u(t)\mathrm{d}t \cos(n\pi x)$$
$$= \sum_{n=0}^{\infty} \int_0^T \Psi_n[t, u(t)]\mathrm{d}t \cos(n\pi x),$$

where

$$\Psi_0(t, u) = u,$$
$$\Psi_n(t, u) = 2(-1)^n \exp\,[-n^2\pi^2(T - t)]u,$$
$$t \in [0, T],\ n = 1, 2, \ldots .$$

Since $g(.) \in L_2(0, 1)$, it possesses a half-range Fourier series,

$$g(x) = \sum_{n=0}^{\infty} \alpha_n \cos(n\pi x).$$

Hence, the control problem reduces to finding a measurable control function

$$u(t) \in [-1, 1],\ t \in [0, T],$$

which satisfies

$$\int_0^T \Psi_n[t, u(t)]\mathrm{d}t = \alpha_n,\ n = 0, 1, 2, \ldots, \tag{VIII.5}$$

and minimizes the functional (VIII.4). We proceed to enlarge the set $V$; we proceed just as in Chapter 1:

(i) For a fixed control function $u(.)$, the mapping

$$F \to \int_0^T F[t, u(t)]\mathrm{d}t,\ F \in C(\Omega),$$

defines a positive linear functional on $C(\Omega)$.

(ii) There exists then a unique positive Radon measure $\mu$ on $\Omega$ such that

$$\int_0^T F[t, u(t)]\mathrm{d}t = \int_\Omega F\mathrm{d}\mu \equiv \mu(F),\ F \in C(\Omega). \tag{VIII.6}$$

(iii) The original minimization problem is replaced by one in which we seek the minimum of $\mu(f_0)$ over a set $Q$ of positive Radon measures on $\Omega$; these measures are required to have certain properties which are abstracted from those satisfied by admissible controls. First, from (VIII.6),

$$|\mu(F)| \leqslant T \sup_{\Omega} |F(t, u)|;$$

hence,

$$\mu(1) = \int_{\Omega} \mathrm{d}\mu \leqslant T.$$

(iv) Also, the measures in $Q$ must satisfy an abstracted version of (VIII.5):

$$\mu(\Psi_n) = \alpha_n, \; n = 0, 1, 2, \ldots .$$

Note that this is possible since the functions

$$\gamma_n \in C(\Omega), \; n = 0, 1, 2 \ldots .$$

Finally, suppose that $G \in C(\Omega)$ does not depend on $u$, that is,

$$G(t, u_1) = G(t, u_2)$$

for all $t \in [0, T]$, $u_1, u_2 \in [-1, 1]$. Then, measures in $Q$ must satisfy

$$\int_{\Omega} G \mathrm{d}\mu = \int_0^T G(t, u) \mathrm{d}t = a_G,$$

where $u$ is an arbitrary number in the set $[-1, 1]$, and $a_G$ is the integral of $G(., u)$ over $[0, T]$, independent of the actual value of $u$.

(v) Therefore, $Q$ is a set of positive Radon measures on $\Omega$ which can be written in the form

$$Q = S_1 \cap S_2 \cap S_3,$$

where

$$S_1 = \{\mu \in \mathscr{M}^+(\Omega) \colon \mu(1) \leqslant T\}$$

$$S_2 = \{\mu \in \mathscr{M}^+(\Omega) \colon \mu(\Psi_n) = \alpha_n, \; n = 0, 1, 2, \ldots\}$$

$$S_3 = \{\mu \in \mathscr{M}^+(\Omega) \colon \mu(G) = a_G, \; G \in C(\Omega) \text{ and independent of } u\}.$$

If we topologize the space $\mathscr{M}(\Omega)$ by the weak*-topology, we can see, much as in the corresponding analysis in Chapter 2, that $Q$ is compact, since $S_1$ is compact, the sets $S_2$ and $S_3$ are intersections of inverse images of closed sets on the real line, and therefore closed. The functional

$$\mu \in Q \to \mu(f_0) \in R \qquad \text{(VIII.7)}$$

is continuous, and thus it achieves a minimum on $Q$. We have shown, thus, the counterpart of Proposition II.2:

**Proposition VIII.1** The measure-theoretical control problem, which consists in finding the minimum of the functional (VIII.7) over the subset $Q$ of $\mathcal{M}^+(\Omega)$, possesses a minimizing solution $\mu^*$, say, a measure in $Q$.

In the previous case we showed that the action of the optimal measure could be approximated by that of an optimal trajectory–control pair; here we must show that it can be approximated by that of a (piecewise constant) control.

With each piecewise constant admissible control $u(.)$ we may associated a measure $\mu_u$ in $\mathcal{M}^+(\Omega) \cap S_1 \cap S_3$. Let $Q_1$ be the set of all such measures $\mu_u$; then Theorem 1 of GHOUILA-HOURI [1] shows that, when $\mathcal{M}(\Omega)$ is given the weak*-topology, $Q_1$ is dense in $\mathcal{M}^+(\Omega) \cap S_1 \cap S_3$. (This theorem suggested our analysis between (IV.3) and Theorem IV.1; as mentioned then, the problem there is harder.) A basis of closed neighbourhoods in this topology is given by sets of the form

$$\{\mu: |\mu(F_n)| \leqslant \varepsilon, n = 1, 2, \ldots, k + 2, \varepsilon > 0\},$$

where $k$ is an integer, $F_n \in C(\Omega)$, $n = 1, 2, \ldots, k + 2$, and $\varepsilon > 0$. In any weak*-neighbourhood of $\mu^*$ (the minimizing measure of Proposition VIII.1), it is therefore possible to find a measure $\mu_u$ corresponding to a piecewise constant control. In particular, we can put

$$F_1 = f_0, F_2 = \Psi_0, \ldots, F_{k+2} = \Psi_k;$$

a piecewise control $u_k(.)$ can therefore be found such that

$$\left| \int_0^T f_0[t, u_k(t)]\mathrm{d}t - \mu^*(f_0) \right| \leqslant \varepsilon$$

$$\left| \int_0^T \Psi_n[t, u_k(t)]\mathrm{d}t - \alpha_n \right| \leqslant \varepsilon, n = 0, 1, \ldots, k.$$

Thus, by employing the control $u_k(.)$, we can get within $\varepsilon$ of the minimum value $\mu^*(f_0)$. The analysis of the relationship between the desired final state $g(.)$ and $y_k(., T)$, $x \in [0, 1]$, the one attained by the use of the control $u_k(.)$, is, however, somewhat more complicated. The Fourier coefficients of $y_k(., T)$ are

$$\beta_n = \int_0^T \Psi_n[t, u_k(t)]\mathrm{d}t, n = 0, 1, \ldots;$$

thus,

$$|\beta_n - \alpha_n| \leqslant \varepsilon, n = 0, 1, \ldots, k.$$

We can show now that if $\varepsilon$ is chosen small enough, and $k$ large enough, the distance between $g(.)$ and $y_k(., T)$ in $L_2(0, 1)$ can be made as small as desired:

**Proposition VIII.2** Given $\delta > 0$, we may choose $k$ and $\varepsilon > 0$ such that

$$\int_0^1 [y(x, T) - g(x)]^2 \mathrm{d}x \leqslant \delta. \tag{VIII.8}$$

*Proof* Since the piecewise constant control $u_k(.)$ has its range in $[-1, 1]$ for all $t \in [0, T]$, the Fourier coefficients $\beta_n$ of $y_k(., T)$ satisfy

$$|\beta_n| \leqslant 2 \int_0^T \exp[-n^2\pi^2(T - t)]|u_k(t)|\mathrm{d}t \leqslant 2/(n\pi)^2, \; n = 1, 2, \ldots. \tag{VIII.9}$$

Similarly, since it is assumed that the desired final state $g(.)$ is reachable with an admissible control, $|\alpha_n|$ satisfies the same inequality as $|\beta_n|$. Thus,

$$\begin{aligned}\int_0^1 [y(x, T) - g(x)]^2 \mathrm{d}x &= \sum_{n=0}^{L} (\beta_n - \alpha_n)^2 + \sum_{n=L+1}^{\infty} (\beta_n - \alpha_n)^2 \\ &\leqslant \sum_{n=0}^{L} (\beta_n - \alpha_n)^2 + 16 \sum_{n=L+1}^{\infty} 1/(n\pi)^4. \end{aligned} \tag{VIII.10}$$

Since the last summation in this expression is the tail of a convergent series, we may choose $L$ such that

$$16 \sum_{n=L+1}^{\infty} 1/(n\pi)^4 \leqslant \delta/2.$$

The integer $k$ can now be chosen as one satisfying

$$k \geqslant \max[L, (1/2\delta) - 1]. \tag{VIII.11}$$

Then,

$$16 \sum_{n=k+1}^{\infty} 1/(n\pi)^4 \leqslant \delta/2; \tag{VIII.12}$$

we choose

$$\varepsilon = \sqrt{[\delta/2(k + 1)]}.$$

From (VIII.11), it follows that $1 + k \geqslant 1/(2\delta)$, from which in turn it follows that $\varepsilon \leqslant \delta$. In the neighbourhood defined by choosing $\varepsilon$ and $k$ as above, a $\mu_u$ exists which corresponds to a piecewise constant control $u(.)$ for which we must have

$$|\beta_n - \alpha_n| \leqslant \varepsilon, \; n = 0, 1, \ldots, k;$$

hence,

$$\sum_{n=0}^{k} (\beta_n - \alpha_n)^2 \leqslant (k + 1)\varepsilon^2 = \delta/2.$$

Combining this last relation with (VIII.10) and (VIII.12) completes the proof of the relation (VIII.1), and then of the proposition. □

Finally, we shall develop some conditions under which the original, classical, unmodified problem has a classical solution. We present this result as an example of what can be achieved in this direction; our philosophy here, as in the rest of the book, is that any sufficiently close approximation to the optimal measure is a solution to the classical problem. As we shall see below, one can guarantee a classical solution to this problem only under convexity conditions on the function $f_0$.

**Proposition VIII.3** Suppose that the function $f_0$ appearing in the performance criterion (VIII.4) satisfies the following conditions: (i) The derivative $f_{0u}$ exists and is uniformly continuous in the interior of $\Omega$. (ii) The function $f_0$ is convex in $u \in [-1, 1]$ for all $t \in [0, T]$.

Then there exists an admissible control $u^*(.)$ such that

$$I[u^*(.)] = \inf_{u(.)\in V} I[u(.)] \equiv \rho.$$

*Proof* Since $f_0 \in C(\Omega)$, the functional $I$ defined in (VIII.4) is bounded below; there exists therefore a sequence of admissible controls, $\{u_i(.)\}$ such that

$$\lim_{i\to\infty} I[u_i(.)] = \rho.$$

Since each control in the sequence $\{u_i(.)\}$ is admissible,

$$\int_0^T u_i(t)^2 \mathrm{d}t \leqslant T.$$

Thus, the $L_2$-norm of the controls in this sequence satisfies $\| u_i(.) \| \leqslant \sqrt{(T)}$, $i = 1, 2, \ldots$. We endow $L_2(0, T)$ with the weak topology, which means that the set $\mathscr{W} = \{u(.): \| u(.) \| \leqslant \sqrt{(T)}\}$ is compact, and $\{u_i(.)\}$ has a weakly-convergent subsequence, which we again denote by $\{u_i(.)\}$, and whose limit we denote by $v(.)$. We claim that $v(.) \in L_2(0, T)$, a result which follows directly from the weak compactness of $\mathscr{W}$; also,

$$\int_0^T \Psi_k[t, v(t)]\mathrm{d}t = \alpha_k, \; k = 0, 1, 2, \ldots,$$

since if this equality were false for some $k$, then an $\varepsilon > 0$ would exist with

$$\int_0^T \exp\,[-k^2\pi^2(T - t)][v(t) - u_i(t)]\mathrm{d}t > \varepsilon$$

for all $i$. However, since $\exp\,(.) \in L_2(0, T)$, this contradicts the fact that $\{u_i(.)\}$ converges weakly to $v(.)$. We also claim that $|v(t)| \leqslant 1$ a.e. on $[0, T]$. Suppose that $|v(t)| > 1$ on some subset of $[0, T]$ having nonzero Lebesgue measure. Let $p(.)$ be the function on $[0, T]$ defined by

$$p(t) = 1 \quad t \in \{s: v(s) > 1\}$$

$$p(t) = -1 \quad t \in \{s: v(s) < -1\}$$

$$p(t) = 0 \quad t \in \{s: |v(s)| \leqslant 1\}.$$

Since $v(.)$ is measurable, $p(.) \in L_2(0, T)$, and

$$\int_0^T p(t)v(t)\mathrm{d}t > \int_0^T p(t)u_i(t)\mathrm{d}t$$

for all $i$. This contradicts the fact that $\{u_i(.)\}$ converges weakly to $v(.)$. We have therefore shown that $v(.)$ is admissible. Consider now the convexity and differentiability assumptions on $f_0$; they imply that

$$f_0(t, v_1) \geqslant f_0(t, v_2) + (v_1 - v_2) f_{0u}(t, v_2)$$

for every $v_1, v_2 \in [-1, 1\}$, $t \in [0, T]$. Hence,

$$\int_0^T f_0[t, u_i(t)]\mathrm{d}t \geqslant \int_0^T f_0[t, v(t)]\mathrm{d}t + \int_0^T [u_i(t) - v(t)] f_{0u}[t, v(t)]\, \mathrm{d}t.$$

Therefore,

$$\begin{aligned} \rho &= \lim_{i\to\infty} I[u_i(.)] \\ &= \lim_{i\to\infty} \int_0^T f_0[t, u_i(t)]\mathrm{d}t \\ &\geqslant I[v(.)] + \lim_{i\to\infty} \int_0^T [u_i(t) - \mathrm{v}(t)] f_{0u}[t, v(t)]\mathrm{d}t. \end{aligned} \qquad \text{(VIII.13)}$$

By assumption, $f_{0u}$ is uniformly continuous in the interior of $\Omega$, and thus bounded on $\Omega$. Since $t$, $v(t)$, $t \in [0, T]$, are both measurable, the function $f_{0u}[., v(.)] \in L_2(0, T)$. Since $\{u_i(.)\}$ converges weakly to $v(.)$, the last limit in (VIII.13) reduces to zero, and therefore

$$\rho \geqslant I[v(.)],$$

which, considering the definition of $\rho$ as the greatest lower bound of the set of all values of the functional $I$ over the set $V$ of admissible controls, implies – considering our result above, that $v(.)$ is admissible – that $I[v(.)] = \rho$. The infimum is thus achieved at $v(.)$, which plays the role of the control $u^*(.)$ of the proposition. □

We have followed in this presentation WILSON and RUBIO [1]. In the following section we study the controllability of the diffusion equation.

## 2 Strong controllability of the diffusion equation

Let us consider again the optimal control problem associated with the diffusion equation (VIII.1) with boundary conditions (VIII.2); it is desired to minimize a functional such as (VIII.4) in which the integrand may be also a function of the state $y(x, t)$, and the integration is then on $[0, 1] \times [0, T]$; the control is assumed to be in $L_2(0, T)$, and no constraints will be imposed on its magnitude. Suppose that the state $g(.) \in L_2(0, 1)$ is not reachable by an admissible control, nor even by a measure – that is, suppose that set of measures $Q$ introduced in the previous section is empty. Then no minimization can be carried out, and the problem has no solution. There are many optimal control problems, even in one-dimensional state spaces, without solution because the desired final state cannot be reached by means of an admissible control. There is, however, something very special about the diffusion equation: the set of states which can be reached by means of controls in $L_2(0, T)$ is dense in $L_2(0, 1)$ (see, for instance, MACCAMY *et al.* [1]). This fact is not particularly helpful if the set of measures $Q$ is empty; but it suggests that maybe we could arrange things so that every state in $L_2(0, 1)$ is reachable from the origin; this would of course entail, by a now familiar argument, enlarging the set of admissible controls, further indeed than before – it is not enough to enlarge it to a set of measures! We must go *beyond measures.*

Let $g(.) \in L_2(0, 1)$ have a half-range Fourier series as in the previous section. Then the solution of (VIII.1) with the boundary conditions (VIII.2) corresponding to a control $u(.) \in L_2\ (0, T)$ satisfies the terminal condition $y(.,T) = g(.)$ in $L_2(0, 1)$ if and only if

$$\int_0^T \Phi_n(t)u(t)\mathrm{d}t = \alpha_n,\ n = 0, 1, 2 \ldots \qquad \text{(VIII.14)}$$

with

$$\Phi_n(t) = \exp[-n^2\pi^2(T - t)],\ t \in [0, T],\ n = 0, 1, 2, \ldots.$$

It should be noted that we have changed the notation from that used in the previous section. It is apparent from (VIII.14) that the problem of attaining a given state $g(.)$ at time $T$ can be studied by considering the *moment problem* (VIII.14). It is not simple to ascertain conditions on the moments $\alpha_n$ so that the problem has a solution. From the results of FATTORINI and RUSSELL [1] it can be shown quite simply that there is a control $u(.) \in L_2(0, T)$ satisfying (VIII.14) if there is a constant $C$ so that the moments $\alpha_n$ satisfy

$$|\alpha_n| \leqslant C \exp(-n^2\pi^2),\ n = 0, 1, 2, \ldots. \qquad \text{(VIII. 15)}$$

It is possible, however, to think of many functions in $L_2(0, 1)$ whose moments do not decrease with $n$ as rapidly as this condition requires.

The result (VIII.15) is obtained by considering the functions $\Phi_n$, $n = 0, 1, 2, \ldots$, as elements of $L_2(0, T)$, and looking for a continuous linear functional in this space so as to satisfy (VIII.14); this functional is of course defined by the control $u(.)$. It should be recognized, however, that the functions $\Phi_n$ are elements of many proper (algebraic) subspaces of $L_2(0, T)$, such as the space of infinitely differentiable functions on $[0, T]$ with derivatives satisfying a boundedness condition (*) below, which, when provided with an appropriate topology, has a dual which contains the space $L_2(0, T)$ as well as other elements. If no control $u(.) \in L_2(0, T)$ exists so as to satisfy (VIII.14), it could be that among these other elements one, or more, can be found so as to provide a solution to the moment problem.

Let $\mathscr{F}$ be the space of real-valued functions infinitely differentiable on $[0, T]$ (that is, these functions have uniformly continuous derivatives on $(0, T)$ of all orders), such that

$$\sup_{[0, T]} |D^j \Phi(t)| \leqslant c L^j, \tag{*}$$

for some constants $c$, $L$ dependent in general on the particular function $\Phi \in \mathscr{F}$. A topology can be put on $\mathscr{F}$ as follows. Let us have $L > 0$, and $\mathscr{F}_L$ the space of infinitely differentiable functions on $[0, T]$ which satisfy the inequality above for this particular $L$ and some $c$, which may depend on $\Phi$. Then

(i) $\mathscr{F}_{L1} \subset \mathscr{F}_{L2}$, $L_1 < L_2$.

(ii) Define on $\mathscr{F}_L$ the norm

$$\|\Phi\|_L = \sup_{j;\, t \in [0, T]} (1/L^j)|D^j \Phi(t)|.$$

With the topology induced by this norm, $\mathscr{F}^L$ is a Banach space.

(iii) For $L_2 > L_1$, the topology induced by $\mathscr{F}_{L1}$ on $\mathscr{F}_{L2}$ is the same as given originally to $\mathscr{F}_{L1}$; the norms $\|\,.\,\|_{L1}$ and $\|\,.\,\|_{L2}$ are equivalent on $\mathscr{F}_{L1}$. Indeed, for all $t \in [0, T]$ all $j$, all $\Phi \in \mathscr{F}_{L1}$,

$$(1/L_2{}^j)|D^j \Phi(t)| \leqslant (1/L_1{}^j)|D^j \Phi(t)|,$$

so that $\|\Phi\|_{L2} \leqslant \|\Phi\|_{L1}$. Therefore, the injection from $\mathscr{F}_{L1}$ into $\mathscr{F}_{L2}$ is continuous and thus an isomorphism into; therefore the two topologies are equivalent.

(iv) The space $\mathscr{F}$ is the union of all spaces $\mathscr{F}_L$:

$$\mathscr{F} = \bigcup_L \mathscr{F}_L$$

It is possible, therefore, to put on $\mathscr{F}$ the LF structure generated by the spaces $\mathscr{F}_L$. We are interested in the dual $\mathscr{S}$ of $\mathscr{F}$, this space with the LF topology; $\mathscr{S}$ is the new space in which we will find a solution to the problem of moments.

**Proposition VIII.4** The linear functional $s$ on $\mathscr{F}$ defined by (VIII.16) is in $\mathscr{S}$ (that is, it is continuous):

$$s(\Phi) = \sum_{k=0}^{\infty} (-1)^k \mu_k (D^k \Phi), \tag{VIII.16}$$

for all $\Phi \in \mathscr{F}$; here $\mu_k$, $k = 0, 1, 2, \ldots$, are Radon measures on $[0, T]$ such that

$$\sum_{k=0}^{\infty} L^k \int_{[0,T]} \mathrm{d}|\mu_k| < \infty, \qquad \text{(VIII.17)}$$

for all $L \geqslant 0$. If $s$ satisfies (VIII.16) for all $\Phi \in \mathscr{F}$, we write

$$s = \sum_{k=0}^{\infty} \mathrm{D}^k \mu_k. \qquad \text{(VIII.18)}$$

*Proof* The functional (VIII.16) is well defined on $\mathscr{F}$. It is continuous if and only if its restriction to $\mathscr{F}_L$ is continuous, for all $L > 0$. Take $\Phi \in \mathscr{F}_L$, for some $L > 0$. Then

$$\begin{aligned} |s(\Phi)| &\leqslant \sum_{k=0}^{\infty} |\mu_k(\mathrm{D}^k\Phi)| \\ &= \sum_{k=0}^{\infty} L^k |\mu_k \mathrm{D}^k(\Phi) L^{-k}| \\ &\leqslant \sum_{k=0}^{\infty} L^k \left[ \int_{[0,T]} \mathrm{d}|\mu_k| \right] \sup_{k,t} L^{-k} |\mathrm{D}^k \Phi| \\ &= \left\{ \sum_{k=0}^{\infty} L^k \int_{[0,T]} \mathrm{d}|\mu_k| \right\} \|\Phi\|_L, \end{aligned}$$

so that the restriction of $s$ to $\mathscr{F}_L$ is continuous. Since this is true for any $L > 0$, $s$ is a continuous functional on $\mathscr{F}$. □

It should be noted that $\mathscr{M}(\Omega)$ is, according to this proposition, a subset of $\mathscr{S}$; also, $L_2(0, T)$ is a subset of $\mathscr{S}$, if we identify a function $u \in L_2(0, T)$ with a Radon measure in the usual manner.

Since the functions $\Phi_n$ are in $\mathscr{F}$ for all $n$, the problem of moments now consists in finding $s \in \mathscr{S}$ such that $s(\Phi_n) = \alpha_n$, $n = 0, 1, 2, \ldots$. We prove an important result:

**Proposition VIII.5** Let $g \in L_2(0, 1)$. Then the set $\mathscr{S}_g \subset \mathscr{S}$: $\mathscr{S}_g = \{s \in \mathscr{S}$: $s(\Phi_n) = \alpha_n$, $n = 0, 1, 2, \ldots$, $s$ is of the form (VIII.16), the measures $\mu_k$, $k = 0, 1, 2, \ldots$, associated with it are atomless$\}$ is nonempty.

*Proof* Consider the associated problem of moments

$$\int_0^T \Phi_n(t) u(t) \mathrm{d}t = a_n \exp(-n^2\pi^2), \; n = 0, 1, 2 \ldots.$$

This problem, according to the results quoted above (see (VIII.15)) has a solution $u(.) \in L_2(0, T)$, since the moments $\alpha_n$ of $g \in L^2(0, 1)$ satisfy $|\alpha_n| \leqslant C$, $n = 0, 1, 2, \ldots$, for some $C > 0$. Then,

$$|\alpha_n \exp(-n^2\pi^2)| \leqslant C \exp(-n^2\pi^2),$$

that is, condition (VIII.15). Define the element $s$ by

$$s(\Phi) = \sum_{k=0}^{\infty} \int_0^T [\mathrm{D}^k \Phi(t)]\,(1/k!)u(t)\mathrm{d}t, \tag{VIII.19}$$

for $\Phi \in \mathscr{F}$. According to Proposition VIII.4, this functional is in $\mathscr{S}$; we have chosen as measures $\mu_k$, $k = 0, 1, 2, \ldots$, the following:

$$\int_{[0,T]} \Phi \mathrm{d}\mu_k \equiv \int_0^T \Phi(t)\,(1/k!)(-1)^k u(t)\mathrm{d}t,\ \Phi \in F.$$

This particular choice of measures satisfies the condition (VIII.17), since for any $L \geqslant 0$,

$$\sum_{k=0}^{\infty} L^k \int_{[0,T]} \mathrm{d}|\mu_k| = \sum_{k=0}^{\infty} L^k (1/k!)(-1)^k \int_0^T |u(t)|\mathrm{d}t$$
$$= \exp(-L) \int_0^T |u(t)|\mathrm{d}t < \infty,$$

since $u(.) \in L_2(0, T) \subset L_1(0, T)$. Therefore, $s \in \mathscr{S}$, is of the form (VIII.16), and the measures $\mu_k$, $k = 0, 1, 2, \ldots$ associated with it are atomless. We can compute the moments $s(\Phi_n)$, since $u(.)$ is a solution of the associated problem of moments,

$$s(\Phi_n) = \sum_{k=0}^{\infty} \int_0^T [\mathrm{D}^k \Phi_n(t)](1/k!)u(t)\mathrm{d}t$$
$$= \left[\sum_{k=0}^{\infty} (n^2\pi^2)^k (1/k!)\right] \int_0^T \Phi_n(t)u(t)\mathrm{d}t$$
$$= \left[\sum_{k=0}^{\infty} (n^2\pi^2)^k (1/k!)\right] \alpha_n \exp(-n^2\pi^2) = \alpha_n,\ n = 0, 1, 2, \ldots.$$

Therefore, the functional $s$ defined by (VIII.19) is in $\mathscr{S}_g$, and this set is nonempty. □

We have been successful in solving, in some sense, the problem of moments associated with the diffusion equation. We now study the possibilities of, first, defining the action of a functional $s \in \mathscr{S}$, not only at the final time $t = T$, but also at intermediate times; and then, of approximating this action by that of a sequence of controls in $L_2(0, T)$.

Suppose that $u(.) \in L_2(0, T)$. Then the corresponding solution of (VIII.1) can be written as follows:

$$y(x, t) = a_0(t; u) + 2\sum_{n=1}^{\infty} a_n(t; u)\,(-1)^n \cos(n\pi x),\ (x, t) \in (0, 1) \times (0, T),$$

with

$$a_0(t; u) = \int_0^t u(\xi)\mathrm{d}\xi,\ a^n(t; u) = \int_0^t \Phi_n(\xi, t)u(\xi)\mathrm{d}\xi,\ n = 1, 2, \ldots. \tag{VIII.20}$$

Consider, for a functional $s \in \mathscr{S}$ of the form (VIII.16), the following expressions:

$$a_0(t; s) = \int_0^t \mathrm{d}\mu_0(\xi),$$

$$a_n(t; s) = \sum_{k=0}^{\infty} (-1)^k \int_0^t \mathrm{D}^k \Phi_n(\xi, t)\mathrm{d}\mu_k(\xi), \; n = 1, 2, \ldots. \quad \text{(VIII.21)}$$

Without addressing ourselves yet to the question of whether the expressions for $a_n(.; s)$, $n = 0, 1, \ldots$, are well defined, we note that, if $s$ is a functional $s_u$ corresponding to a control $u \in L_2(0, T)$, then the expressions (VIII.21) become identical with those in (VIII.20); the new expressions are true extensions of the previous ones. We prove that the functions $a_n(.; s)$ are well defined and show some fundamental properties of these functions.

**Proposition VIII.6** The functions $a_n(.; s)$, $n = 0, 1, \ldots$, defined in (VIII.21), are well defined and continuous on $[0, T]$.

*Proof* Let $K$ be a positive integer, and let $s_K$ be the functional

$$s_K(\Phi) = \sum_{k=0}^{K} (-1)^K \mu_k(\mathrm{D}^k \Phi), \; \Phi \in \mathscr{F}.$$

Then, for any $n \geqslant 0$,

$$\begin{aligned} a_n(t; s_K) &= \sum_{k=0}^{K} (-1)^k \int_0^t \mathrm{D}^k \, \Phi_n(\xi, t)\mathrm{d}\mu_k(\xi) \\ &= \sum_{k=0}^{K} (-1)^k (\pi n)^{2k} \int_0^t \Phi_n(\xi, t)\mathrm{d}\mu_k(\xi), \; t \in [0, T]. \quad \text{(VIII.22)} \end{aligned}$$

This function is continuous, for any $n$ and $K$, since the measures $\mu_K$ are atomless. Let $C[0, T]$ be the space of continuous functions on $[0, T]$ with the topology of uniform convergence. Fix the integer $n \geqslant 0$, and consider the sequence $\{a_n(.; s_k), K = 1, 2, \ldots)$ of functions in $C[0, T]$. We show that this is a Cauchy sequence. Indeed,

$$\begin{aligned} &\sup_{t \in [0, T]} |a_n(t; s_{K1}) - a_n(t; s_{K2})| \\ &= \sup_t \left| \sum_{k=K_1+1}^{K_2} (-1)^k (\pi n)^{2k} \int_0^t \Phi_n(\xi, t)\mathrm{d}\mu_k(\xi) \right| \\ &\leqslant \sup_t \sum_{k=K_1+1}^{K_2} (\pi n)^{2k} \int_0^t \Phi_n(\xi, t)\mathrm{d}|\mu_k(\xi)| \\ &\leqslant \sum_{k=K_1+1}^{K_2} (\pi n)^{2k} \int_0^T \mathrm{d}|\mu_k(\xi)| \xrightarrow[K_1, K_2 \to \infty]{} 0, \end{aligned}$$

since the series

$$\sum_{k=0}^{\infty} (\pi n)^{2k} \int_0^T \mathrm{d}|\mu_k(\xi)|$$

converges, according to (VIII.17) and the fact that $\Phi_n \in \mathscr{F}$. Without loss of generality, we have assumed that $K_2 > K_1$. It follows that the sequence $\{a_n(.; s_K)\}$ converges uniformly to a continuous function $a_n(.; s)$. The convergence depends on the index $n$. □

This proof already contains the germ of an approximation scheme: it is possible to approximate the action of the functional $s$ by that of the truncated functional $s_K$, at every $t \in [0, T]$. Let us develop this scheme further. Define $\omega = (-1, T + 1)$, and let $C_c^{\infty}(\omega)$ be the space of infinitely differentiable functions with compact support in $\omega$, with the LF topology. For any $\Phi \in C_c^{\infty}(\omega)$ and a fixed integer $K > 0$, we define the functional $L_t^K$: $C_c^{\infty}(\omega) \to R$ by

$$\langle L_t^K, \Phi \rangle = \sum_{k=0}^{K} (-1)^k \int_0^t \mathrm{D}^k \Phi(\xi) \mathrm{d}\mu_k(\xi), \qquad \text{(VIII.23)}$$

for any $t \in [0, T]$. We prove

**Proposition VIII.7** The functional $L_t^K$ defined by (VIII.23) is continuous, that is, in $\mathrm{D}'(\omega)$. There is a sequence $\{u_j^K\}$ of functions in $L_2(\omega)$ such that

$$u_j^K \to L_t^K \text{ in } \mathrm{D}'(\omega) \textit{ strongly}. \qquad \text{(VIII.24)}$$

*Proof* Indeed, if $\tau$ is a compact subset of $\omega$, and the support of $\Phi$ is in $\tau$,

$$|\langle L_t^K, \Phi \rangle| \leqslant K \sup_{0 \leqslant k \leqslant K} \int_0^t \mathrm{d}|\mu_k| \sup_{0 \leqslant k \leqslant K} \sup_{\tau} |\mathrm{D}^k \Phi(t)|,$$

which implies that $L_t^K$ is in $\mathrm{D}'(\omega)$. The existence of the sequence $\{u_j^K\}$ follows from some standard results from distribution theory (see TRÈVES [1], p. 302). □

It is necessary to explore the possibility of approximating the action of the family of functionals $L_t^K$, $0 \leqslant t < T$, by the corresponding restrictions of $u_j^K$. Let $\omega_t \equiv (0, t)$. Then, we have

**Proposition VIII.8** Let $u_{jt}^K$ be the restriction of $u_j^K$ to $\omega_t$ for $j = 1, 2, \ldots$. Then,

$$u_{jt}^K \to L_t^K \text{ in } \mathrm{D}'(\omega) \text{ strongly}. \qquad \text{(VIII.25)}$$

*Proof* (i) We consider the functional $\Lambda_t^K$, the restriction of $L_T^K$ to $\omega_t$. This functional is in $\mathrm{D}'(\omega_t)$. The mapping

$$L_T{}^K \to \Lambda_t{}^K,$$

of $D'(\omega)$ into $D'(\omega_t)$, is continuous when both these spaces have the strong topology.

(ii) It follows from (VIII.24) and (i) that the restriction of $u_j{}^K$ to $\omega_t$, to be called $u_{jt}{}^K$, satisfies

$$u_{jt}{}^K \to \Lambda_t{}^K \quad \text{in } D'(\omega_t) \text{ strongly.} \tag{VIII.26}$$

(iii) The functional $\Lambda_t{}^K$ can be extended to $C_c{}^\infty(\omega)$, its extension being the functional $L_t{}^K$ defined in (VIII.23). We prove that (VIII.26) implies that

$$\langle u_{jt}{}^K, \Phi\rangle \to \langle L_t{}^K, \Phi\rangle, \; \Phi \in C_c{}^\infty(\omega); \tag{VIII.27}$$

since weak and strong convergence are equivalent for sequences of distributions, this result implies the contention (VIII.25).

(iv) The restriction of functions $\Phi \in C_c{}^\infty(\omega)$ to $\omega_t$ are in $C^\infty(\omega_t)$. We shall prove that (VIII.26) implies that

$$\langle u_{jt}{}^K, \Psi\rangle \to \langle L_t{}^K, \Psi\rangle, \; \Psi \in C^\infty(\omega_t); \tag{VIII.28}$$

in particular, this implies (VIII.27). Indeed, if $\Psi \in C^\infty(\omega_t)$, there exists a sequence $\{\Phi_i\}$ of functions in $C_c{}^\infty(\omega_t)$ which tend to $\Psi$ in the $C_c{}^\infty(\omega_t)$-topology. We have the following diagram:

$$\begin{array}{ccc} \langle u_{jt}{}^K, \Phi_i\rangle & \xrightarrow[j]{} & \langle L_t{}^k, \Phi_i\rangle \\ \Big\downarrow i & & \Big\downarrow i \\ \langle u_{jt}{}^k, \Psi\rangle & \xrightarrow[j]{?} & \langle L_t{}^k, \Psi\rangle \end{array} \tag{VIII.29}$$

Indeed,

$$\begin{aligned} |\langle u_{jt}{}^K, \Psi\rangle - \langle L_t{}^K, \Psi\rangle| \leqslant {} & |\langle u_{jt}{}^K, \Psi - \Phi_i\rangle| + |\langle L_t{}^K, \Psi - \Phi_i\rangle| \\ & + |\langle u_{jt}{}^K, \Phi_i\rangle - \langle L_t{}^K, \Phi_i\rangle|. \end{aligned} \tag{VIII.30}$$

The last term in this inequality holds the clue to the proof of convergence. The set $\{\Phi_i\}$ is bounded in $C^\infty(\omega_t)$; since

$$u_{jt}{}^K \to L_t{}^K$$

strongly in $D'(\omega)$, the convergence of

$$\langle u_{jt}{}^K, \Phi_i\rangle - \langle L_t{}^K, \Phi_i\rangle \to 0$$

does not depend on the index $i$. Given $\varepsilon > 0$, there is then an integer $M_1(\varepsilon, K, t)$ so that the last term in (VIII.30) is less than $\varepsilon/3$ if $j > M_1$. We choose $j$ in this way, and then we choose $i$ so that both the first and second terms are less than $\varepsilon/3$. The contention (VIII.28) follows, and then (VIII.27), then (VIII.25). □

In order to apply this result to our problem of approximation, we must extend the functions $\xi \to \Phi_n(\xi, t)$ to $\omega$. The resultant set of functions has an interesting property:

**Proposition VIII.9** Let $N$ be a positive integer, and let $B$ be the set of all functions of the form

$$\xi \to \Phi_n{}^*(\xi, t) = \hat{\Phi}_n(\xi, t)\gamma(\xi), \ \xi \in \omega, \ n = 0, 1, 2, \ldots,$$

and any $t \in [0, T]$; $\hat{\Phi}_n$ is the extension of $\Phi_n$ to $\omega$ given by the same expression following (VIII.14), and $\gamma \in C_c{}^\infty(\omega)$ is such that $\gamma(t) = 1$ on $[0, T]$. Then the set $B$ is bounded in $C_c{}^\infty(\omega)$.

*Proof* Note that, according to our definition, the functions $\Phi_n{}^*(., t)$, $n = 0, 1, 2, \ldots$, are in $C_c{}^\infty(\omega)$ and $\Phi_n{}^*(\xi, t) = \Phi_n(\xi, t)$, $\xi \in [0, T]$. Since the function $\gamma$ has compact support, all the functions in $B$ have their support in a compact subset $\omega'$ of $\omega$. Given $m > 0$, we must show that the set

$$\left\{\sup_{p \leqslant m} \sup_{\xi \in \omega'} |\mathrm{D}^p \Phi(\xi)|, \ \Phi \in B\right\} \tag{VIII.31}$$

is bounded in the real line. Indeed, for any

$$p \leqslant m, \ 0 \leqslant n \leqslant N, \ t \in [0, T], \ \xi \in \omega',$$

we have

$$|\mathrm{D}^p \Phi_n(\xi, t)| = (n\pi)^{2p} \exp[-n^2\pi^2(t - \xi)| \leqslant (N\pi)^2 \exp[N^2\pi^2(2 + T)];$$

thus the set (VIII.31) is bounded in the real line, and $B$ is bounded in $C_c{}^\infty(\omega)$. □

According to this result and Proposition VIII.8,

$$\sup_{t \in [0, T]} \sup_{1 \leqslant n \leqslant N} |\langle u_{jt}{}^K - L_t{}^K, \Phi_n(., t)\rangle| \to 0$$

as $j \to \infty$. This expression is equivalent to

$$\int_0^t u_j{}^K(\xi)\Phi_n(\xi, t)\mathrm{d}\xi \to a_n(t; s_K), \ n = 0, 1, 2, \ldots, N, \tag{VIII.32}$$

*uniformly on* $[0, T]$. Of course, the convergence depends on the indices $N$ and $K$. We have been successful in that we can approximate uniformly the action of $s_K$ at every $t \in [0, T]$ on a finite, but arbitrarily large, number of functions $\xi \to \Phi_n(\xi, t)$. Finally, from this result we can derive the final approximation scheme.

**Proposition VIII.10** Given an integer $N > 0$ and any $\varepsilon > 0$, there exists a control $u \in L_2(0, T)$ such that

$$\sup_{t\in[0,T]} \sup_{0\leqslant n\leqslant N} |a_n(t; u) - a_n(t; s)| < \varepsilon.$$

In particular,

$$|\alpha_n - a_n(T; u)| < \varepsilon, n = 0, 1, 2, \ldots, N.$$

*Proof* For fixed $N$, the proof of Proposition VIII.6 implies that there is an integer $K(N)$ such that

$$\sup_{t\in[0,T]} \sup_{0\leqslant n\leqslant N} |a_n(t; s_K) - a_n(t; s)| < \varepsilon/2,$$

for all $K > K(N)$. Take an integer $K_1$ satisfying this inequality. Then (VIII.31) implies that

$$\sup_{t\in[0,T]} \sup_{0\leqslant n\leqslant N} |a_n(t; u_{jt}^{K1}) - a_n(t; s_{K1})| < \varepsilon/2,$$

for all $j$ higher than an integer $j(K_1)$. The proposition follows by putting

$$u = u_j^{K1}, j > j(K_1). \qquad \square$$

We have been successful in approximating the action of $s$, in the sense that any finite number of functions $a_n(.; s)$, $n = 0, 1, 2, \ldots$, associated with the functional $s$ may be uniformly approximated on $[0, T]$ to within any accuracy by using a control function from $L_2(0, T)$. We note that in order to be quite sure that any state in $L_2(0, 1)$ can be reached at time $T$, it is not enough to enlarge the space of admissible controls to a space of measures; we must go beyond, and indeed beyond the conventional distributions, to an even larger space. In this way, the framework is set for the solution of control problems involving the diffusion equation and fixed terminal states.

## References

In Section 1, we followed WILSON and RUBIO [1]. In Section 2, RUBIO and WILSON ([1]–[2]).

# Appendix
# Functions, functionals and measures

## 1 Spaces of functions. Linear continuous functionals

We have considered in this book spaces of functions and measures defined on the set $\Omega$, a closed subset of a finite-dimensional euclidean space, which is therefore locally compact and Hausdorff. (See SCHUBERT [1] for a treatment of these concepts.) We shall present in this appendix a summary of the necessary background material, in a somewhat more general framework than is strictly necessary; we shall consider spaces of functions and measures on a topological space, to be denoted also by $\Omega$, which is locally compact and Hausdorff but otherwise arbitrary. The reader with no interest in the extra generality thus gained – gained with little extra work – can consider the space $\Omega$ as being the product set $J \times A \times U$ appearing in the main body of the text. A warning: the theory to be developed in this appendix does not, in general, apply to the case considered in Chapter 7, when the state space is a Hilbert space; in that case, the set $\Omega$ is not locally compact.

Let, then, $\Omega$ be a locally compact, Hausdorff topological space. We shall consider in this section the space $C(\Omega)$ of all continuous real-valued functions on $\Omega$, with a topology to be defined below, and then its dual $C(\Omega)'$. We assume the reader to be familiar with the theory of locally convex linear topological spaces, in particular with the concept of a basis of continuous seminorms defining a topology. (See TRÈVES [1], Chapter 7.)

As a matter of fact, we shall endow $C(\Omega)$ with the (locally convex) topology defined by the seminorms

$$F \to p_K(F) \equiv \sup_{z \in K} |F(z)|,\ F \in C(\Omega), \tag{A.1}$$

where $K$ is any compact subset of the space $\Omega$. If $\Omega$ is compact Hausdorff – which is the situation encountered in most of this book – then $C(\Omega)$ is normable, and it becomes a Banach space with topology defined by the norm

$$F \to \|F\| = \sup_{\Omega} |F(z)|,\ F \in C(\Omega); \tag{A.2}$$

we note that this topology is equivalent to that generated by the seminorms defined in (A.1) since $p_K(F) \leqslant p_\Omega(F) = \|F\|$, for all $F \in C(\Omega)$ and all compact subsets $K$ of $\Omega$.

We shall study in Section 3 continuous linear functionals on $C(\Omega)$ in the case when $\Omega$ is not necessarily compact, after we have established a connection between functionals and classical measures; otherwise we would have to employ some topological arguments which are beyond the scope of this appendix.

If $\Omega$ is compact Hausdorff, the continuous linear functionals on $C(\Omega)$ – this space with the topology induced by the norm (A.2) – are those linear functions $\Lambda$: $C(\Omega) \to R$ such that

$$|\Lambda(F)| \leqslant \gamma \| F \|, f \in C(\Omega); \tag{A.3}$$

here $\gamma$ is a constant independent of the particular function $F \in C(\Omega)$, but which will in general depend on the particular linear functional $\Lambda$. As a matter of fact, we can define a norm in the space $C(\Omega)'$ of all continuous linear functionals on $C(\Omega)$, even if we shall not in general have much use for the corresponding induced topology. Let $\Gamma$ be the set of all constants $\gamma$ such that, for a fixed linear continuous functional $\Lambda$, the expression (A.3) is true. Then,

$$\| \Lambda \| = \inf \Gamma = \sup_{|F| \neq 0} |\Lambda(F)| \Big/ \| F \| \tag{A.4}$$

is a norm in $C(\Omega)'$.

A functional in $C(\Omega)'$ is known as a *Radon measure*. If the functional is *positive*, that is, if

$$F(z) \geqslant 0 \text{ on } \Omega \Rightarrow \Lambda(F) \geqslant 0,$$

then it is known as a *positive Radon measure*. The use of the word *measure* seems to suggest a connection between these functionals and the measures of classical measure theory. We shall start preparing the ground now for establishing this connection later on, in Section 3. We define a function, totally specified by a *positive* functional $\Lambda \in C(\Omega)'$, which assigns to each subset of the space $\Omega$ a real number. Let $P(\Omega)$ be the *power set* of the space $\Omega$, that is, the set of all its subsets, and $O(\Omega)$ the set of all open subsets of $\Omega$. Then, abusing somewhat the notation, given a positive functional $\Lambda \in C(\Omega)'$ we define a function $\Lambda$: $O(\Omega) \to R$ by

$$\Lambda(G) = \sup \{\Lambda(F): 0 \leqslant F(z) \leqslant \chi_G, z \in \Omega, F \in C(\Omega)\}, G \in O(\Omega), \tag{A.5}$$

where $\chi_G$ is the characteristic function of the open set $G$. Note that this function is generally not continuous, so that the functional $\Lambda$ cannot be applied directly to it; the expression (A.5) represents a sort of extension of this functional, to characteristic functions of open sets. Note also that (A.5) implies that $\Lambda(\varnothing) = 0$; $\varnothing$ is the empty set. Further, we define a new set function, $\Lambda^*$: $P(\Omega) \to R$ by

$$\Lambda^*(A) = \inf \{\Lambda(G): G \supset A, G \in O(\Omega)\}. \tag{A.6}$$

We can now prove several properties of the function $\Lambda^*$, which will eventually, in Section 3, help us associate a classical measure with the positive linear functional $\Lambda$.

**Proposition A.1** Let $\Lambda$ be a positive functional on $C(\Omega)$, with $\Omega$ a compact Hausdorff topological space. The function $\Lambda^*$ defined in (A.6) is an *outer measure* on $P(\Omega)$; that is,

(i) $\Lambda^*(\emptyset) = 0$,

(ii) $A \subset B,\ A, B \in P(\Omega) \Rightarrow \Lambda^*(A) \leqslant \Lambda^*(B)$,

(iii) $A = \bigcup_{j=1}^{\infty} A_j,\ A_j \in P(\Omega), j = 1, 2, \ldots \Rightarrow \Lambda^*(A) \leqslant \sum_{j=1}^{\infty} \Lambda^*(A_j)$.

*Proof* (a) The assertion (i) follows from the fact that $\Lambda(\emptyset) = 0$.

(b) If $G_1$ and $G_2$, $G_1 \subset G_2$, are open subsets of $\Omega$, then

$$0 \leqslant \Lambda(G_1) \leqslant \Lambda(G_2),$$

since

$$\{F: 0 \leqslant F(z) \leqslant \chi_{G1}\} \subset \{F: 0 \leqslant F(z) \leqslant \chi_{G2}\}.$$

Thus, the assertion (ii) follows.

(c) Consider a sequence $\{G_i, i = 1, 2, \ldots\}$ of open sets in $\Omega$. Then,

$$\Lambda\left(\bigcup_{i=1}^{\infty} G_i\right) \leqslant \sum_{i=0}^{\infty} \Lambda(G_i). \tag{A.7}$$

Indeed, consider just two sets, $G_1$ and $G_2$, and let $\chi_1$ and $\chi_2$ be their respective characteristic functions. Let $\varepsilon > 0$; according to the definition (A.5) we can find a function $F \in C(\Omega)$ such that $\Lambda(F) > \Lambda(G_1 \cup G_2) - \varepsilon$. Define now a closed set $E = \{z \in \Omega: F(z) \geqslant \varepsilon\}$, which is then a subset of $G_1 \cup G_2$. It is possible to find closed sets $E_1 \subset G_1$, $E_2 \subset G_2$ and functions $F_1$ and $F_2$ such that $F_i(z) = 1$ on $E_i$ and $0 \leqslant F_i(z) \leqslant \chi_i(z)$, $z \in \Omega$, $i = 1, 2$. Then $F(z) < F_1(z) + F_2(z) + \varepsilon$ on $\Omega$. Thus,

$$\Lambda(G_1 \cup G_2) - \varepsilon < \Lambda(F) \leqslant \Lambda(F_1) + \Lambda(F_2) + \varepsilon\Lambda(1),$$

where 1 is the function equal to 1 everywhere on $\Omega$; we have used the fact that the functional $\Lambda$ is positive. Since this functional is continuous, $\Lambda(1) \leqslant \|\Lambda\|$; since $\varepsilon$ is arbitrary, we have, finally, that

$$\Lambda(G_1 \cup G_2) \leqslant \Lambda(G_1) + \Lambda(G_2).$$

The proof of the corresponding inequality for a finite number of sets $\{G_i, i = 1, 2, \ldots, n\}$ follows by a process of induction, while (A.7) then follows by taking limits.

(d) We can now prove the assertion (iii). Let $G_j \supset A_j$, $G_j \in O(\Omega)$, $j = 1, 2, \ldots$, such that $\Lambda(G_j) < \Lambda^*(A_j) + \varepsilon/2^j$, with $\varepsilon > 0$ and otherwise arbitrary; definition (A.6) guarantees that such a sequence of open sets does exist. Then $A$ is contained in the union of all the sets $G_j$, $j = 1, 2, \ldots$, and

$$\Lambda^*(A) \leqslant \Lambda^*\left(\bigcup_{j=1}^{\infty} G_j\right) = \Lambda\left(\bigcup_{j=1}^{\infty} G_j\right) \leqslant \sum_{j=1}^{\infty} \Lambda(G_j) \leqslant \sum_{j=1}^{\infty} \Lambda^*(A_j) + \varepsilon,$$

which, since $\varepsilon$ is arbitrary, implies the condition (iii). □

In the next section we summarize briefly some aspects of classical measure theory, so as to be able to establish in Section 3 a connection between Radon and classical measures. This connection is, simply, that any Radon measure can be identified with a classical measure on $\Omega$; we shall only establish this fact for positive measures.

## 2 A summary of classical measure theory

A *measure space* is a triple $(\Omega, \mathscr{S}, \mu)$, where:

(i) $\Omega$ is a set, not necessarily provided with a topology. We are, however, only interested in the situation in which this set is the same locally compact Hausdorff topological space considered in most of the first section. Some of the more interesting results in classical measure theory spring from the interaction between topological and purely measure-theoretical properties of the spaces of functions associated with $\Omega$.

(ii) $\mathscr{S}$ is a *σ-algebra* of subsets of $\Omega$, that is, a nonempty collection of subsets of $\Omega$ which satisfies the following conditions:
(a) $A \in \mathscr{S} \Rightarrow \Omega \backslash A \in \mathscr{S}$.

(b) $A_1, \ldots, A_n, \ldots \in \mathscr{S} \Rightarrow \bigcup_{n=1}^{\infty} A_n \in \mathscr{S}$

Note that some consequences of these properties are: $\Omega$ and the empty set $\varnothing$ are in $\mathscr{S}$; if $A_1, \ldots, A_n, \ldots$, are in $\mathscr{S}$, then $\bigcap_{n=1}^{\infty} A_n$ is in $\mathscr{S}$.

Let $\mathscr{E}$ be an arbitrary collection of subsets of $\Omega$. A $\sigma$-algebra $\mathscr{S}$ is said to be the *σ-algebra generated* by $\mathscr{E}$ if:
(a) $\mathscr{S}$ contains $\mathscr{E}$.
(b) If $\mathscr{W}$ is a $\sigma$-algebra and $\mathscr{W}$ contains $\mathscr{E}$, then $\mathscr{W}$ contains $\mathscr{S}$.

The $\sigma$-algebra $\mathscr{B}$ generated by the collection $O(\Omega)$ of all open subsets of the topological space $\Omega$ will be of great importance in subsequent developments. The sets in this $\sigma$-algebra are known as *Borel sets*.

If a set A $\in \mathscr{S}$, we shall say that it is *measurable*; if necessary, we say that it is *$\mathscr{S}$-measurable*.

(iii) $\mu$ is a *measure*, that is, a mapping of the $\sigma$-algebra $\mathscr{S}$ into the extended real number system, $\mu: \mathscr{S} \to R \cup \pm\{\infty\}$, which satisfies the following conditions:

(a) $\mu(\varnothing) = 0$;

(b) It is *countably additive*, that is,

$$\mu\left(\bigcup_{i=1}^{\infty} A_i\right) = \sum_{i=1}^{\infty} \mu(A_i),$$

where $\{A_i, i = 1, 2, \ldots\}$ is a sequence of disjoint sets in $\mathscr{S}$.

If $\mathscr{S} = \mathscr{B}$, the $\sigma$-algebra generated by the open sets in $\Omega$, then the measure $\mu$ is said to be a *Borel* measure provided that the measure of every compact set in $\Omega$ is finite. A Borel measure is *regular* if for every Borel set $A$

$$\mu(A) = \inf\{\mu(G), A \subset G, G \in O(\Omega)\}; \tag{A.8}$$

that is, a Borel measure is regular if its action on Borel sets can be evaluated with any degree of accuracy by its action on the open sets.

If $P(z)$ is a proposition about $z \in \Omega$ which is true for all $z$ but for a set of measure 0, then $P(z)$ is said to be *true almost everywhere* (a.e), or *true $\mu$-almost everywhere.*

If $\mu(A) \geqslant 0$ for all $A \in \mathscr{S}$, we shall say that the measure $\mu$ is *positive*; otherwise, it is said to be *signed.*

Our main interest concerning measure spaces consists in the integration of functions defined on the set $\Omega$ with respect to the measure $\mu$. Let $(\Omega, \mathscr{S}, \mu)$ be a measure space, $F$ a function defined on $\Omega$ with values in the extended real line, and $A$ a measurable set. The function $F$ is said to be *measurable on $A$* if for any real number $a$ the set

$$\{z \in A: F(z) > a\} = F^{-1}\{(a, \infty]\}$$

is measurable, that is, it is in the $\sigma$-algebra $\mathscr{S}$. If $A = \Omega$ then the funcion is said to be *measurable*. Many results, most quite simple, can be established in connection with this concept: any linear combination of measurable functions is measurable; if $F$ is measurable $|F|$ is measurable; if $F$, $H$ are measurable the product function $FH$ is measurable; if $F$ is measurable $F^+$ and $F^-$ are measurable; if the functions in a sequence $\{F_n\}$ are all measurable then the function defined by taking the pointwise supremum over this sequence is measurable; a set $A$ is measurable if and only if its characteristic function $K_A$ is measurable.

Let $F$ be a nonnegative function measurable on a set $A$. Further, let $\{A_i, i = 1, 2, \ldots, n\}$ be a finite sequence of sets such that

(i) $A_i \cap A_j = \varnothing$, $i \neq j$; that is, the sets in the sequence are disjoint.

(ii) $A_i$ is measurable, $i = 1, 2, \ldots, n$.

(iii) $\mathrm{A} = \bigcup_{i=1}^{n} A_i$.

It is said that $\{A_i\}$ is a *partition* of the set $A$. Further, let

$$s = \sum_{i=1}^{n} \{\inf_{A_i} F(z)\} \mu(A_i),$$

which is well defined since the function $F$ is nonnegative everywhere. Finally, define the integral of the function $F$ over the measurable set $A$ with respect to the measure $\mu$ as

$$\int_A F\mathrm{d}\mu \equiv \sup s \geqslant 0, \qquad \text{(A.9)}$$

where the supremum of the numbers $s$ is taken over all possible partitions of the set $A$ satisfying the conditions (i), (ii) and (iii) above. Of course, the integral in (A.9) can take the value $\infty$; if it is finite, the function $F$ is said to be *integrable* – or *summable* – over the set $A$ with respect to the measure $\mu$.

If the function $F$ is not necessarily nonnegative, its integral can be defined formally as

$$\int_A F\mathrm{d}\mu = \int_A F^+\mathrm{d}\mu - \int_A (-F^-)\mathrm{d}\mu. \qquad \text{(A.10)}$$

Of course, this definition only makes sense for those functions for which at least one of the integrals in the right-hand side of (A.10) is finite. In such a case, if

$$-\infty < \int_A F\mathrm{d}\mu < \infty,$$

the function $F$ is said to be integrable, or summable.

These integrals have many properties; we summarize below only the most important for our purposes. In all cases, $(\Omega, \mathcal{S}, \mu)$ is a measure space.

(i) If $F_1$, $F_2$ are nonnegative functions which are measurable on a set $A$,

$$\int_A (F_1 + F_2)\mathrm{d}\mu = \int_A F_1\mathrm{d}\mu + \int_A F_2\mathrm{d}\mu.$$

(ii) If $F$ is nonnegative on a set $A = A_1 \cup A_2$, with $A_1$ and $A_2$ measurable and disjoint (that is, $A_1 \cap A_2 = \varnothing$), then

$$\int_A F\mathrm{d}\mu = \int_{A_1} F\mathrm{d}\mu + \int_A F\mathrm{d}\mu$$

(iii) Consider an increasing sequence of nonnegative functions $\{F_i\}$, which are all measurable on a set $A$. Suppose, further, that this sequence converges pointwise to a function $F$ on $A$. Then,

$$\lim_{i\to\infty} \int_A F_i\mathrm{d}\mu = \int_A F\mathrm{d}\mu.$$

(iv) Consider a sequence of nonnegative functions $\{F_i\}$, all measurable on a set $A$. Then

$$\int_A \underline{\lim}\, F_i \mathrm{d}\mu \leqslant \underline{\lim} \int_A F_i \mathrm{d}\mu.$$

This result is known as *Fatou's lemma.*

(v) Consider a sequence of functions $\{F_i\}$, all measurable on a set $A$, and such that $|F_i(z)| \leqslant H(z)$, for almost all $z \in A$; $H$ is a summable function. If this sequence converges pointwise to a function $F$ a.e. on $A$, then

$$\int_A F\mathrm{d}\mu = \lim_{i \to \infty} \int_A F_i \mathrm{d}\mu.$$

This result is known as the *dominated convergence theorem.*

(vi) Suppose that the functions $F_1$ and $F_2$ are integrable on a set $A$. Then the function $\alpha_1 F_1 + \alpha_2 F_2$ is integrable for all real $\alpha_1$ and $\alpha_2$, and

$$\int_A (\alpha_1 F_1 + \alpha_2 F_2)\mathrm{d}\mu = \alpha_1 \int_A F_1 \mathrm{d}\mu + \alpha_2 \int_A F_2 \mathrm{d}\mu.$$

We shall establish in the next section a relationship between regular Borel measures and Radon measures, when the space $\Omega$ is compact; the case when this space is only locally compact (and Hausdorff) will also be considered.

## 3 Functionals and measures

We shall first suppose that the space $\Omega$ is compact and Hausdorff, and prove a theorem which in a more general version – for arbitrary continuous functionals and signed measures – is known as Riesz's theorem; we do not need all this generality in this book, because our linear functionals – and then our measures – are, without exception, positive.

**Theorem A.1** Let the space $\Omega$ be compact Haudorff, and $\Lambda$ a positive continuous functional on $C(\Omega)$, that is, a positive Radon measure. Then there is a unique positive regular Borel measure $\mu$, on $\Omega$, such that

$$\Lambda(F) = \int_\Omega F\mathrm{d}\mu, \ F \in C(\Omega). \tag{A.11}$$

*Proof* (i) We go back first to the framework of outer measures of Proposition A.1 and related material. We say that a subset $B$ of the space $\Omega$ is $\Lambda^*$*-measurable* if for every subset $A$ of $\Omega$,

$$\Lambda^*(A) = \Lambda^*(A \cap B) + \Lambda^*(A \backslash B). \tag{A.12}$$

We show that all Borel sets in $\Omega$ are $\Lambda^*$-measurable. Indeed, let $A$ be a subset of $\Omega$, $G$ an open set in $\Omega$, and $G_1$ an open set in $\Omega$ such that $A \subset G_1$. Put $C_1 \equiv G \cap G_1$, $C_2 \equiv G_1 \backslash C_1 = G_1 \cap G^c$, with $G^c = \Omega \backslash G$, the complement of $G$ with respect to the space $\Omega$. Then we can find functions $F_1$ and $F_2$ such that

$$\Lambda(F_i) > \Lambda(C_i) - \varepsilon, \; i = 1, 2,$$

for any $\varepsilon > 0$. Together with the fact that $\Lambda$ is a linear functional, the definition of the outer measure $\Lambda^*$ implies that,

$$\Lambda(F_1 + F_2) = \Lambda(F_1) + \Lambda(F_2) > \Lambda(C_1) + \Lambda(C_2) - 2\varepsilon = \Lambda^*(C_1) + \Lambda^*(C_2) - 2\varepsilon.$$

Also, since $C_1 \cup C_2 = G_1$, $\Lambda(G_1) \geqslant \Lambda(F_1 + F_2)$; since $\Lambda^*(A) = \inf \Lambda(G)$ and

$$\Lambda^*(G \cap G_1) \geqslant \Lambda^*(G \cap A), \; \Lambda^*(G^c \cap G_1) \geqslant \Lambda^*(G^c \cap A),$$

then,

$$\Lambda^*(A) > \Lambda^*(G \cap A) + \Lambda^*(G^c \cap A) - 2\varepsilon,$$

which, since $\varepsilon > 0$ is otherwise arbitrary,

$$\Lambda^*(A) \geqslant \Lambda^*(G \cap A) + \Lambda^*(G^c \cap A), \tag{A.13}$$

which implies (A.12) with $G = B$, since $\Lambda^*(A)$ is never higher than the right-hand side of (A.13). Since the $\Lambda^*$-measurable subsets of the space $\Omega$ form a $\sigma$-algebra, the equality (A.13) is actually valid for all Borel sets $B$, which are then all $\Lambda^*$-measurable.

(ii) The restriction of the outer measure $\Lambda^*$ to the $\sigma$-algebra composed of the Borel sets is a (positive Borel) measure, that is, it is countably additive. We must show that it is regular. Indeed, denote this measure by $\mu$; then, if $A$ is any Borel set,

$$\mu(A) = \Lambda^*(A) = \inf\{\Lambda(G): A \subset G, G \in O(\Omega)\},$$

which is the same as the expression (A.8) by means of which we defined the regularity of a Borel measure.

(iii) Let $\varepsilon > 0$. There exists $F \in C(\Omega)$ such that $0 \leqslant F(z) \leqslant 1$ on $\Omega$, and $\Lambda(F) > \|\Lambda\| - \varepsilon$. Thus, $\|\Lambda\| - \varepsilon < \Lambda(F) \leqslant \Lambda(1) \leqslant \|\Lambda\|$; since this is true for all $\varepsilon > 0$,

$$\Lambda(1) = \|\Lambda\| = \Lambda(\Omega). \tag{A.14}$$

Thus, the measure of any compact set in $\Omega$ is finite, another requirement for $\Lambda$ to be a Borel measure.

(iv) We now show that the measure $\mu$ introduced above satisfies the main contention of the theorem, expression (A.11). Suppose first that the function $F$ satisfies $0 \leqslant F(z) \leqslant 1$ on $\Omega$. We choose an integer $n \geqslant 1$, and let

$$G_i = \{z \in \Omega: F(z) > i/n\}, \; i = 0, 1, 2, \ldots, n.$$

Then

$$\varnothing = G_n \subset \ldots \subset G_1 \subset G_0.$$

Let $F_i(z) = 0$ if $F(z) < (i - 1)/n$, $F_i(z) = n[F(z) - (i - 1)/n]$ if $(i - 1)/n \leqslant F(z) \leqslant i/n$, $F_i(z) = 1$ if $F(z) > i/n$, for $i = 1, 2, \ldots, n$. Then, writing $\chi_i$ for the characteristic function of the set $G_i$, $i = 1, 2, \ldots n$,

$$0 \leqslant F_i(z) \leqslant \chi_i,$$

$$(1/n) \sum_{i=1}^{n} F_i(z) = F(z),\ z \in \Omega,$$

and then

$$\begin{aligned} \Lambda(F) &= (1/n) \sum_{i=1}^{n} \Lambda(F_i) \\ &\leqslant (1/n) \sum_{i=1}^{n} \mu(G_{i-1}) \\ &= \sum_{i=1}^{n} \mu(G_{i-1} \backslash G_i)\,(i - 1)/n + (1/n)\mu(\Omega); \end{aligned}$$

since on $G_{i-1} \backslash G_i$

$$(i - 1)/i < F(z) \leqslant i/n,$$

then

$$\Lambda(F) \leqslant \int_\Omega F\mathrm{d}\mu + (1/n)\mu(\Omega),$$

for $n = 1, 2, \ldots$; thus,

$$\Lambda(F) \leqslant \int_\Omega F\mathrm{d}\mu, \tag{A.15}$$

since $\mu(\Omega)$is finite; see (A.14). The same argument can be applied to $1 - F$, which implies that $\Lambda(F)$ cannot be less than the right-hand side of (A.15); thus

$$\Lambda(F) = \int_\Omega F\mathrm{d}\mu, \tag{A.16}$$

This equality does hold true for any function in $C(\Omega)$, without the qualification that $0 \leqslant F(z) \leqslant 1$. If $F$ is any nonnegative function in $\Omega$, a positive constant $\alpha$ can be found so that the function $\alpha F$ satisfies this constraint, from which (A.16) follows readily for $F$. If $F$ is any function in $C(\Omega)$, a constant $\beta$ can be found so that $F' \equiv F + \beta$ is nonnegative on $\Omega$. Thus,

$$\Lambda(F') = \Lambda(F) + \beta\Lambda(1) = \int_\Omega F'\mathrm{d}\mu = \int_\Omega F\mathrm{d}\mu + \beta\int_\Omega \mathrm{d}\mu,$$

which implies that $F$ satisfies (A.16), since this same equality implies that

$$\Lambda(1) = \int_\Omega d\mu.$$

(v) Finally we consider the uniqueness of the measure $\mu$ constructed above. Suppose that $\upsilon$ is another positive regular Borel measure such that

$$\Lambda(F) = \int_\Omega F d\nu, \; F \in C(\Omega).$$

Then

$$\int_\Omega F d(\nu - \mu) = 0, \; F \in C(\Omega),$$

which implies $\mu = \nu$, since these are regular measures. (See FREMLIN [1], 73D). □

We shall identify Radon and regular Borel measures in this book; it is sometimes more convenient to visualize a measure as a functional, other times better to see it as a set function. We say that the measure $\mu$ of this theorem is a *representing measure* for the functional $\Lambda$, or that *it represents* this functional.

We shall now prove that continuous positive functionals on $C(\Omega)$ have representing measures even when $\Omega$ is not compact. Let this be the case, and consider a positive measure on $\Omega$. This measure is said *to be zero on an open set* $G$ in $\Omega$ if $\mu(F) = 0$ for all $F \in C(\Omega)$ whose support is contained in $G$; here, as in the rest of this section, we shall use $\mu(F)$ as shorthand for the integral of $F$ on $\Omega$ with respect to the measure $\mu$. The *support* of the measure $\mu$ is defined to be the complement with respect to $\Omega$ of the union of all open sets $G$ on which $\mu$ is zero. We can prove our main representation theorem for spaces $\Omega$ which are not necessarily compact.

**Theorem A.2** If $\mu$ is a positive measure on a locally compact Hausdorff space $\Omega$ whose support is compact, then the mapping $F \to \mu(F)$, $F \in C(\Omega)$, is a continuous positive linear functional. If $\Lambda$ is a positive linear functional on $C(\Omega)$, then there exists a unique measure $\mu$ on $\Omega$ whose support is compact such that $\Lambda(F) = \mu(F)$, $F \in C(\Omega)$.

*Proof* The first part of the theorem is simple to prove. The mapping $F \to \mu(F)$ is linear and positive because of the properties of the integral with respect to the measure $\mu$. It is also continuous, since

$$|\mu(F)| = \left| \int_\Omega F d\mu \right| \leqslant \sup_K |F(z)| \int_K d\mu,$$

where $K$ is the (compact) support of the measure $\mu$. Considering that the topology of the space $C(\Omega)$ is defined in this case by the family of seminorms in (A.1), the contention – that the mapping $F \to \mu(F)$ is continuous – follows from this inequality.

If $\Lambda$ is a continuous linear form on $C(\Omega)$, there must be a seminorm from the family defined by (A.1) such that

$$|\Lambda(F)| \leqslant p_k(F), F \in C(\Omega);$$

this implies that the support of the functional $\Lambda$ is in $K$ – that is, that functions in $C(\Omega)$ whose support is outside $K$ are mapped into 0 by $\Lambda$. The space $C(K)$ of restrictions of the functions in $C(\Omega)$ to the compact set $K$ is *normed* by the seminorm $p_k$. Thus, the restriction of the functional $\Lambda$ to $C(K)$ can be represented, by Theorem A.1, by a measure whose support is contained in $K$, and is therefore compact. This measure is such that $\Lambda(F) = \mu(F), F \in C(\Omega)$. □

The main results of this section, Theorems A.1 and A.2, are also valid for signed measures and functionals. In the next section we study the extremal points of a set of measures.

## 4 Extremal points and the minimum of a linear form

We consider first a linear space $X$, not endowed with a topology. If $x_1$ and $x_2$ are in $X$, the *line segment* between these points is the set

$$\{x \in X: x = \theta x_1 + (1 - \theta)x_2, 0 \leqslant \theta \leqslant 1\}.$$

An *open segment* between these points is the set

$$\{x \in X: x = \theta x_1 + (1 - \theta)x_2, 0 < \theta < 1\},$$

that is, the line segment between the same points minus its endpoints $x_1$ and $x_2$.

A nonempty set $C \subset X$ is *convex* if for each pair of points in $C$ the line segment between them is in $C$.

A point $x$ of a compact subset $C$ of the linear space $X$ is said to be *nonextremal* if there is an open segment contained wholly in $C$ which contains the point $x$. A point of $C$ which is not nonextremal is said to be *extremal.* A point $x$ of $C$ is extremal if and only if

$$x = (x_1 + x_2)/2, x_1, x_2 \in X \Rightarrow x_1 = x_2 = x.$$

The interplay between these properties of sets and points and a topology on $X$ which is compatible with the linear structure of this space makes this a very attractive area of mathematics. Our interest in these matters is, in this book, concerned only with the space of measures $\mathcal{M}(\Omega)$ with the weak*-topology, with a basis of neighbourhoods of zero of $\mathcal{M}(\Omega)$ given by

$$U_\varepsilon = \{\mu \in \mathcal{M}(\Omega): |\mu(F_j)| < \varepsilon, j = 1, 2, \ldots, r\},$$

for every $\varepsilon > 0$ and every finite subset $\{F_j, j = 1, 2, \ldots, r\}$, $F_j \in C(\Omega)$.

We shall first prove a theorem of existence of extremal points.

**Theorem A.3** Let $\Omega$ be a compact Hausdorff topological space. If $C$ is a weakly*-compact subset of the space of measures $\mathcal{M}(\Omega)$, then it has at least one extremal point.

*Proof* (i) We first define a particular class of subsets of the space $C$. A set $E \subset C$ is said to be *extremal* if

$$\mu_1, \mu_2 \in C, \theta\mu_1 + (1 - \theta)\mu_2 \in E \text{ for some } \theta, 0 < \theta < 1 \Rightarrow \mu_1, \mu_2 \in E.$$

It is possible to partially order the class of all compact extremal subsets of $C$ by set inclusion. Then, noting that $C$ is itself an extremal set – so that the class of extremal sets is nonempty – Zorn's lemma indicates that there is a minimal element, $E_0$, in this class.

(ii) We show that $E_0$ cannot contain more than one point; then, it contains only one point which must be an extremal point of $C$, from which the theorem follows. Indeed, suppose that there are two distinct points, $\mu$ and $\upsilon$, in $E_0$; then there is a function $F \in C(\Omega)$ so that

$$\mu(F) \neq \nu(F).$$

Therefore, the set of measures in $E_0$ at which the function $\mu \to \mu(F)$ attains its minimum is a proper compact subset of $E_0$, which cannot therefore be a minimal element. Thus there is only one point in $E_0$. □

We are in a position to consider linear optimization problems on compact convex sets of measures. We assume from now on that the space $\Omega$ is compact Hausdorff.

**Theorem A.4** Let $C$ be a weakly*-compact convex set in $\mathcal{M}(\Omega)$, and $F \in C(\Omega)$. Then the linear function $\mu \to \mu(F)$ attains its minimum at an extremal point of $C$.

*Proof* Since the function $\mu \to \mu(F)$ is weakly*-continuous, it attains its minimum on the weakly*-compact set $C$. Let $m = \min \mu(F)$ over the set $C$, and define

$$A(a) = \{\mu \in C: \mu(f) \leqslant a\} \subset C.$$

If $a > m$, the set $A(a)$ is weakly*-closed, and therefore weakly*-compact, and nonempty. Define, further,

$$A(m) = \bigcap_{a>m} A(a);$$

the function $\mu \to \mu(f)$ attains its minimum $m$ at every point of this set, which is nonempty, compact and convex; it has an extremal point, $\mu^*$. We shall prove that this is also an extremal point of $C$, which will establish the theorem. Indeed, assume that $\mu^*$ is not an extremal point of $C$. Then there are two measures, $\mu_1$ and $\mu_2$ in $C$ such that

$$\mu^* = \theta\mu_1 + (1 - \theta)\mu_2, \tag{A.17}$$

for some $\theta$ between 0 and 1, and $\mu_1 \neq \mu^*$, $\mu_2 \neq \mu^*$. Then,

$$m = \mu^*(f) = \theta\mu_1(f) + (1 - \theta)\mu_2(f) \geqslant \theta m + (1 - \theta)m = m,$$

so that $\mu_1(f) = \mu_2(f) = m$; that is, $\mu_1$ and $\mu_2$ belong to the set $A(m)$. This contradicts (A.17), since $\mu^*$ is an extremal point of this set. □

Finally, we must obtain an explicit expression for the extremal points of a compact convex set of measures such as the set $C$ in this theorem. It is remarkable that such an expression can be obtained quite simply; we follow ROSENBLOOM [1] throughout.

We start by defining a particular class of measure. We say that a measure in $\mathscr{M}(\Omega)$ is a *prime measure* if it is not the zero measure and it takes values 0 and 1 only. Let $A$ be a Borel set, and $z$ a point in the space $\Omega$. The prime measure which assigns the value 1 to the set $A$ if $z \in A$, and a value zero otherwise, will be called a *unitary atomic measure*, and denoted by $\delta(z)$. We show now that integration with respect to a prime measure is equivalent to integration with respect to a unitary atomic measure.

**Proposition A.2** Let $\Omega$ be a compact Hausdorff topological space, and $\mu$ a prime measure in $\mathscr{M}(\Omega)$. Then there is a unique point $z$ in $\Omega$ such that

$$\int_\Omega F \mathrm{d}\mu = \int_\Omega F \mathrm{d}\delta(z) = F(z), \; F \in C(\Omega).$$

*Proof* Let $\mathscr{F}$ be the set of all Borel sets $A$ in $\Omega$ such that $\mu(A) = 0$. (This set is a *prime ideal* of the $\sigma$-algebra $\mathscr{B}$ of Borel sets.) Suppose that to each point $x \in \Omega$ one can associate an open set $G(x)$ in $\mathscr{F}$ which contains $x$. Therefore, it is possible to cover $\Omega$ by a finite number of open sets, $G(x_1), \ldots, G(x_n)$, and

$$\mu(\Omega) = \mu \bigcup_{j=1}^{n} G(x_j) \leqslant \sum_{j=1}^{n} \mu[G(x_j)] = 0,$$

which is impossible, since the measure $\mu$ is not identically equal to zero. There is therefore a point $z$ in $\Omega$ such that no open set $G$ which contains $z$ is in the prime ideal $\mathscr{F}$.

Let now $F$ be any function in $C(\Omega)$, and choose a real number $\varepsilon > 0$. There is an open set $G$ which contains $z$ – and which, therefore, is not in $\mathscr{F}$, so that

$\mu(G) \neq 0$ – such that

$$|F(x) - F(z)| \leqslant \varepsilon, \ x \in G, \ F \in C(\Omega).$$

Thus, noting that since $\mu(G) \neq 0$ it must equal 1, since $\mu$ is prime,

$$F(z) - \varepsilon = [F(z) - \varepsilon]\mu(G) \leqslant \int_{\Omega} F d\mu \leqslant [F(z) + \varepsilon]\mu(G) = F(z) + \varepsilon,$$

from which the main statement of the proposition follows. If $\Omega$ is Hausdorff, it is normal (see SCHUBERT [1]), and therefore, if $z_1 \neq z$, there is a continuous function $F$ on $\Omega$ such that $F(z) = 0$, $F(z_1) = 1$, so that

$$0 = \int_{\Omega} F d\mu \neq F(x_1).$$

The uniqueness of the point $z$ is therefore proved. □

The following proposition will be important when we establish the structural results concerning the extreme points of a compact convex set of measures.

**Proposition A.3** Let $\mu$ be a measure in $\mathscr{M}(\Omega)$. If no more than $k$ disjoint Borel sets $A_1, \ldots, A_k$ can be found such that $\mu(A_i) = 0$, $i = 1, 2, \ldots, k$, then $\mu$ is a linear combination of at most $k$ prime measures.

*Proof* Suppose that there are $k$ disjoint Borel sets $A_i$, $i = 1, 2, \ldots, k$, so that

$$\mu(A_i) \equiv \alpha_i \neq 0, \ i = 1, 2, \ldots, k,$$

and define, for any Borel set $A$,

$$\mu_i(A) = \mu(A \cap A_i)/\alpha_i, \ i = 1, 2, \ldots, k.$$

The measures $\mu_i$, $i = 1, 2, \ldots, k$, are prime. Indeed, if $\mu_i(A)$ is neither equal to zero nor to one, then $A$ is nonempty, and the $(k + 1)$ Borel sets

$$A_i, A_1, \ldots, A_{i-1}, A_i \cap A, A_i \backslash A, A_{i+1}, \ldots, A_k$$

are disjoint; also,

$$\mu(A_i \cap A) = \alpha_i \mu(A) \neq 0,$$

while

$$\mu(A_i \backslash A) = \alpha_i \mu_i(\Omega \backslash A) = \alpha_i[1 - \mu_i(A)] \neq 0,$$

which contradicts our supposition, that there are no more than $k$ disjoint Borel sets with nonzero $\mu$-measure.

Let $A$ be an arbitrary Borel set. Then the $(k + 1)$ Borel sets

$$A_1, \ldots, A_k, A \backslash \bigcup_{i=1}^{k} A_i$$

are disjoint, so that

$$\mu\left(A\setminus\bigcup_{i=1}^{k} A_i\right) = 0,$$

which implies

$$\mu(A) = \sum_{i=1}^{k} \mu(A \cap A_i) + \mu\left(A\setminus\bigcup_{i=1}^{k} A_i\right) = \sum_{i=1}^{k} \alpha_i\mu_i(A),$$

which proves the proposition, since the measures $\mu_i = 1, 2, \ldots, k$, are prime. □

Finally, we develop the structural results sought.

**Theorem A.5** Let $F_1, \ldots, F_n$ be continuous functions on the compact Hausdorff topological space $\Omega$, with $F_1(z) = 1$, $z \in \Omega$, and let $Q$ be the set of positive Radon measures on $\Omega$:

$$Q = \{\mu \in \mathcal{M}^+(\Omega)\colon \mu(F_i) = c_i,\ i = 1, 2, \ldots, n\},$$

with $c_1 > 0$. Then if $Q$ is nonempty, it is a compact convex subset of $\mathcal{M}^+(\Omega)$; if $\mu^*$ is an extreme point of $Q$, it is of the form

$$\mu^* = \sum_{i=1}^{n} \alpha_i \delta(z_i),$$

with $\alpha_i \geqslant 0$, $z_i \in \Omega$, $i = 1, 2, \ldots, n$.

*Proof* The proof that $Q$ is compact and convex can be found in Proposition II.2. Let $\mu^*$ be an extremal point of the set $Q$. Suppose there are $(n + 1)$ disjoint Borel sets $A_0, \ldots, A_n$ such that $\mu^*(A_i) \neq 0$, $i = 0, 1, \ldots, n$. Put

$$A_{n+1} = \Omega\setminus(A_1 \cup \ldots \cup A_n) \supset A_0,$$

so that $\mu^*(A_{n+1}) > 0$. To each point $(\alpha_1, \ldots, \alpha_{n+1})$ of the affine space of $(n + 1)$ dimensions we can associate a measure $\mu$ such that

$$\mu(A) = \sum_{i=1}^{n+1} \alpha_i\mu^*(A \cap A_i),$$

for each Borel set $A$. Let $K$ be the set of points in affine space such that the corresponding measures belong to the set $Q$. The point in $K$ corresponding to the extreme point $\mu^*$ is an extremal point of $K$; as such, at most $n$ of the numbers $\alpha_i$, $i = 1, 2, \ldots, n + 1$, are positive; at least one of the them is zero. (See GASS [1], Chapter 3.) Propositions A.3 and A.2 establish then our structural result, the main contention of the theorem. □

**References**

Measure theory: CHOQUET [1], FREMLIN [1], FRIEDMAN [1], HALMOS [1].
Extremal points, convexity: ALFSEN [1], CHOQUET [1], HOLMES [1], ROSENBLOOM [1].

# References


E. M. Alfsen [1] *Compact Convex Sets and Boundary Integrals*. Berlin, Heidelberg and New York: Springer Verlag (1971).

C. Berge [1] *Topological Spaces*. Edinburgh and London: Oliver and Boyd (1963).

N. Bourbaki [1] *General Topology*. New York and London: Academic Press (1966).

H. S. Carslaw [1] *Introduction to the Theory of Fourier's Series and Integrals*. New York: Dover Publications (1930).

L. Cesari [1] Existence theorems for weak and usual optimal solutions in Lagrange problems with unilateral constraints – II: Existence theorems for weak solutions. *Transactions of the American Mathematical Society* **124**, 413–430 (1966).

G. Choquet [1] *Lectures on Analysis*. New York: Benjamin (1969). [2] *Topology*. New York and London: Academic Press (1966).

E. A. Coddington and N. Levinson [1] *Theory of Ordinary Differential Equations*. New York: McGraw-Hill (1955).

G. de Rham [1] *Variétés Différentiables*. Paris: Hermann (1960).

G. di Pillo and L. Grippo [1] A computing algorithm for the application of the epsilon method to identification and optimal control problems. *Ricerce di Automatica* **3**, 54–77 (1972).

R. J. Duffin [1] Infinite programs. *Linear Inequalities and Related Systems*. Edited by H. W. Khun and A. W. Tucker. Princeton: Princeton University Press (1956), pp. 157–70.

R. E. Edwards [1] *Functional Analysis. Theory and Applications*. New York: Holt, Rinehart and Winston (1965).

H. O. Fattorini and D. L. Russell [1] Exact controllability theorems for linear parabolic equations in one space dimension. *Archive for Rational Mechanics and Analysis* **4**, 272–92 (1971).

D. H. Fremlin [1] *Topological Riesz Spaces and Measure Theory*. Cambridge: Cambridge University Press (1974).

A. Friedman [1] *Foundations of Modern Analysis*. New York: Holt, Rinehart and Winston (1970).

S. I. Gass [1] *Linear Programming. Methods and Applications*. 4th Edition. New York: McGraw-Hill (1975).

A. Ghouila-Houri [1] Sur la généralization de la notion de commande d'un système guidable. *Revue Française d'Automatique, Informatique, et Recherche Operationelle* No. 4, 7–32 (1967).

P. Halmos [1] Measure Theory. Princeton: Van Nostrand (1950).

R. B. Holmes [1] *Geometric Functional Analysis and its Applications*. Berlin, Heidelberg and New York: Springer Verlag (1975).

A. D. Ioffe and V. M. Tihomirov [1] *Theory of Extremal Problems*. Amsterdam: North Holland (1979).

S. Lefschetz [1] *Differential Equations: Geometric Theory*, New York: Interscience, a division of John Wiley and Sons (1962).

G. G. Lorentz [1] *Approximation of Functions*. New York: Holt, Rinehart and Winston (1966).

R. C. MacCamy, V. J. Mizel and T. Siedman [1] Realization of temperature distributions. *Journal of Mathematical Analysis and its Applications* **23**, 699–703, (1968).

P. C. Rosenbloom [1] Quelques classes de problèmes extrémaux. *Bulletin de la Société Mathématique de France* **80**, 183–216 (1952).

J. E. Rubio [1] Generalized curves and extremal points. *SIAM Journal on Control* **13**, 28–47 (1975). [2] Extremal points in the calculus of variations. *Bulletin of the London Mathematical Society* **7**, 159–65 (1975). [3] An existence theorem in the calculus of variations. *Journal of Optimization Theory and its Applications* **19**, 81–7 (1976). [4] Extremal points and optimal control theory. *Annali di Matematica* **109**, 165–76 (1976). [5] An existence theorem for control problems in Hilbert spaces. *Bulletin of the London Mathematical Society*. **9**, 70–8 (1977). [6] On optimal control problems in Hilbert spaces: the case of the unbounded controls. *Proceedings of the IMA Conference on Recent Theoretical Developments in Control, Leicester, England, 1976*. Edited by M. Gregson. London: Academic Press (1978), pp. 241–54. [7] Existence and approximation in control problems in Hilbert spaces. *Journal of Mathematical Analysis and its Applications* **69**, 419–27 (1979). [8] The solution of nonlinear optimal control problems in Hilbert spaces by means of linear programming techniques. *Journal of Optimization Theory and its Applications* **30**, 643–61 (1980). [9] Controllability and extensions. *Journal of the Franklin Institute* **309**, 643–61 (1980). [10] Weak global controllability of nonlinear systems. *Journal of Optimization Theory and its Applications* **39**, 251–9 (1983).

J. E. Rubio and D. A. Wilson [1] On a problem of moments of control theory. *Journal of the Franklin Institute* **299**, 291–5 (1975). [2] On the strong controllability of the diffusion equation. *Journal of Optimization Theory and its Applications* **23**, 607–16 (1977).

H. H. Schaefer [1] *Topological Vector Spaces*. New York: Macmillan (1966).

H. Schubert [1] *Topology*. Boston: Allyn and Bacon (1968).

L. Schwartz [1] *Théorie des Distributions*. Paris: Hermann (1966).

F. Trèves [1] *Topological Vector Spaces, Distributions and Kernels*. New York and London: Academic Press (1967).

V. S. Varadarajan [1] Measures on topological spaces. *Transactions of the American Mathematical Society* **48**, 161–228 (1965).

A. M. Vershik [1] Some remarks on the infinite-dimensional problems of linear programming. *Russian Mathematical Surveys* **25**, 117–24 (1971).

J. Warga [1] *Optimal Control of Differential and Functional Equations*. New York and London: Academic Press (1972).

D. A. Wilson and J. E. Rubio [1] Existence of optimal controls for the diffusion equation. *Journal of Optimization Theory and its Applications* **22**, 91–100 (1977).

L. C. Young [1] *Calculus of Variations and Optimal Control Theory*. Philadelphia: W. B. Saunders (1969).

# Index